# Applied Biomechanics
## Second Edition

## Mark D. Ricard, PhD
### University of Texas Arlington

**Library of Congress Cataloging-in-Publication Data**

Ricard, Mark D. 1956–
Applied Biomechanics / Mark D. Ricard. -- 2nd ed.
   p. ; cm.
Includes bibliographical references and index.
ISBN–13: 978-1492223580
ISBN–10: 1492223581
1. Kinesiology. 2. Human mechanics. I. Title.
[DNLM: 1. Biomechanics.  2. Movement--physiology.  3. Musculoskeletal Physiology.  WE  103  2011]
QP303.H354 2013
612.7'6–dc22

ISBN–13: 978-1492223580
ISBN–10: 1492223581

To request permission, please contact

Biomech Publications
Mansfield, TX 76063

email at biomechpub@gmail.com.

# DEDICATION

This book is dedicated in memory of
Anthony Barksdale II.

## Acknowledgements

Over the last 25 years, the multitude of questions raised by students have made significant contributions to the usefulness of this book as a tool of learning.

I wish to personally thank Susie K. Chung for her editorial work in this book. Her perseverance and attention to detail significantly improved the clarity of this book.

*Mark D. Ricard, PhD*

# Table of Contents

# Chapter 1

## INTRODUCTION TO BIOMECHANICS

### STUDENT LEARNING OUTCOMES

**After reading Chapter 1, you should be able to**

- Define biomechanics.
- List the scientists who contributed to the field of biomechanics and briefly describe their contributions.
- Describe the instrumentation used in the field of biomechanics.
- Define mechanics.

### WHAT IS BIOMECHANICS?

Biomechanics is the study of forces and their effects on living systems. The name biomechanics is derived from the study of biological tissues and the field of mechanics. Mechanics can be defined as the study of the effects of forces and torques on the motion of bodies. Figure 1.1 shows the relationship between biomechanics and the fields of anatomy and physiology and the field of mechanics. Biomechanics merges anatomy and physiology with the field of mechanics. The branch of mechanics can be further subdivided into kinematics and kinetics. **Kinematics** is the description of motion which uses position, velocity, and acceleration to

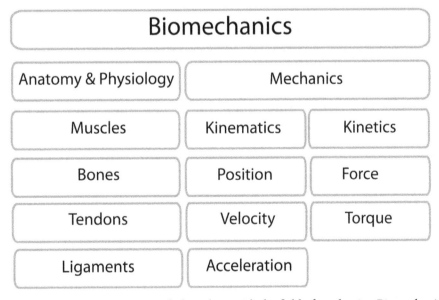

**Figure 1.1** Biomechanics merges anatomy and physiology with the field of mechanics. Biomechanics is the study of forces and their effects on living systems. The name biomechanics is derived from the study of biological tissues and the field of mechanics.

describe motion. **Kinetics is** the cause of motion. In kinetics, forces and torques are analyzed to determine their potential effects on motion.

## RELATIONSHIP OF BIOMECHANICS TO OTHER MOVEMENT DISCIPLINES

The field of biomechanics evolved from the field of kinesiology in the early 1980s. Kinesiology is the study of movement which was originally constrained to studying movement based upon muscles and the motion they produced. Gradually kinesiologists starting using 8 mm and 16 mm film to study motion. When using film, kinesiologists were able to make basic calculations by scaling the film image to life-size dimensions and used calculus, physics and mechanics to analyze the physics of human motion. In the 1980s, the kinesiologists started using computers to analyze their film data. With the advent of computers, the name biomechanics slowly replaced the term kinesiology to describe the study of human movement. Figure 1.2 gives examples of fields of study that now use biomechanics.

Biomechanics is used:
- by coaches to analyze sport movements to improve performance.
- by pedagogy (physical education) professionals to give the learner anatomically and mechanically correct feedback and teaching cues.
- by athletic trainers, physical therapist and sports medicine professionals to analyze injury mechanisms and rehabilitation exercises.
- by ergonomic professionals to analyze work place motions to improve productivity and reduce the likelihood of injury on the job.
- by strength coaches to design exercise programs to improve muscular strength, endurance, and power.

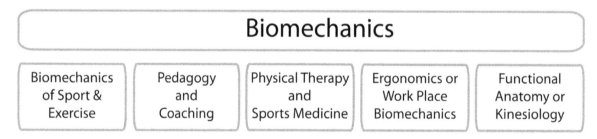

**Figure 1.2** Today biomechanics is used by coaches, teachers, athletic trainers, physical therapists, work place biomechanists, and functional anatomists.

## A BRIEF HISTORY OF BIOMECHANICS AND KINESIOLOGY

Biomechanics derives its roots from the fields of mechanics and biology. Like all sciences, the history of biomechanics begins with the ancient Greeks and progresses forward from that point. The scientists listed below made significant and lasting contributions to biomechanics.

## Aristotle - The Father of Kinesiology

Kinesiology is defined as the science of movement. Kinesiology is derived from the Greek words 'kinein', which means **to move**, and 'logos', which is a **reasoned discourse**. Aristotle (384 - 322 BC) is often referred to as the "father of kinesiology". He analyzed walking and used geometry to describe the actions of muscles (Martin, 1999). Aristotle wrote two books devoted to movement: *De Incessu Animalium* (On the Gait of Animals) and *De Motu Animalium* (On the Motion of Animals).

## Leonardo da Vinci - Mechanical Principles and Anatomy

The next biomechanist of interest is the artist and engineer Leonardo da Vinci (1452 - 1519), who used mechanical principles to study anatomy (Hawking, 2002; Martin, 1999). He used the components of force vectors derived from muscle lines of actions to determine the effects of muscle forces on joint motion. Using his skill as an artist and his knowledge of mechanics, da Vinci drew remarkably detailed drawings of bones, muscles, and tendons.

## Galileo Galilei - The Father of Mechanics

While Aristotle is referred to as the father of kinesiology, Galileo Galilei (1564 - 1642) is often referred to as the father of mechanics. He used mathematics to describe friction forces and ballistics. Galileo's work contributed to the advancement of the scientific method (Hawking, 2002; Martin, 1999). Based upon his observations of moving objects, he proposed his "Principle of Inertia", which is similar to Newton's "Law of inertia". In addition, Galileo's observations of planetary motion enabled him to confirm that Copernicus' hypothesis of a heliocentric solar system was correct.

## Giovanni Borelli - The Father of Biomechanics

Giovanni Alfonso Borelli (1608 - 1679) was an Italian physiologist, physicist, and mathematician who is often described as the "father of biomechanics" (Martin, 1999; Pope, 2005). He studied mathematics under a former student of Galileo. Borelli was the first to describe the action of muscles as levers. He calculated the forces in various joints required to keep the body in equilibrium. Using mathematics and mechanics, Borelli demonstrated that muscle forces are much higher than the gravitational forces they must overcome due to the much smaller moment arms of muscles about a joint center. In addition to his contribution to the study of muscle lever arms, Borelli also determined the position of the center of gravity in humans. Borelli's book *De Motu Animalium* was published shortly after his death. He named is book the same as Aristotle's book as a tribute to Aristotle.

## Isaac Newton - The Laws of Motion

In his book *Principia Mathematica Philosphiae Naturalis*, Isaac Newton (1642 - 1727) formulated the three laws of motion: I. Law of Inertia, II. Law of Acceleration, and III. Law of Reaction (Hawking, 2002; Martin, 1999; Newton, 1687). Newton's principles describe acceleration, deceleration, and inertial movement. His contributions to science include the invention of calculus, a mathematical explanation of accelerated motion, and the theory of gravity.

## Eadweard Muybridge - Stop-Motion Photography

The English photographer Eadweard Muybridge (1830 - 1904) contributed to the development of biomechanics with his photographic studies of human and animal motion (Haas, 1976). He used multiple cameras to capture motion in stop action photographs. Muybridge's human motion studies used multiple images of humans walking, climbing stairs, and boxing. Using his stop-motion technique, Muybridge was able to demonstrate that there is a brief moment where a galloping horse has all four hooves off the ground at once. The human eye can discern 10-12 images per second and is therefore is unable to verify the flight phase for a galloping horse.

## Braune and Fischer - Gait Analysis and Anthropometrics

Christian Wilhelm Braune (1831 - 1892) and Otto Fischer (1861 - 1917) developed the methodology that is used today for gait analysis. They published two papers on gait: free walking and walking with a load (Braune & Fischer 1895, Braune & Fischer 1904). Braune and Fischer contributed greatly to our understanding of how joint forces developed by muscles result in motion. In addition, Fischer was instrumental in the study of the center of gravity in human segments. He froze cadaver segments and then determined the location of the segments' center of gravity about the longitudinal, sagittal, and frontal axes.

## Charles Sherrington - Postural Control and Sensory Feedback Signals

Charles Sherrington (1857 - 1952) made significant contributions to the study of posture and movement (Stuart, 2005; Molnar et al., 2010). Sherrington also advanced our understanding of convergence of descending command signals and sensory feedback signals upon the motor unit.

## A. V. Hill - Muscle Force-Velocity Relationship

The British physiologist Archibald Vivian Hill (1886 - 1977) is best known for his work on the heat of shortening, which led to quantifying the muscle force–velocity relationship (Hill, 1938). Hill was not terribly fond of his given name and preferred to be addressed as A. V. Hill (Van der Kloot, 2011). While A.V. Hill began his undergraduate study in the field of mathematics, he later went on to study physiology, physics and chemistry. Using his background in mathematics, physics, and chemistry, Hill approached the study of muscle from the perspective of a machine. He determined the thermodynamic properties of muscle by measuring the heat produced during muscular contractions. In 1910, A. V. Hill began conducting experiments on heat production in the frog sartorius muscle (Hill, 1910). In his first paper on muscle contraction, he reported a constant relationship between muscle force and the heat developed in muscle for an isometric twitch. In 1919, Hill and Hartree improved their ability to measure the heat produced in a muscle contraction. They designed an instrument called a thermopile, that when connected to a galvanometer was capable of measuring the heat produced by muscular contraction to an accuracy of less than 0.000001° C (Hill, et al.,

1920). A.V. Hill received the Nobel Prize for Physiology and Medicine in 1922 for his discovery related to the production of heat in muscle. Hill shared the prize with the German biochemist Otto Meyerhof.

A.V. Hill is credited with discovering the relationship between muscle force and the velocity of muscle shortening. He measured the velocity of shortening and the change in force as a function of velocity using a maximally stimulated frog sartorius muscle (Hill, 1938). In these experiments, Hill measured the rate of heat production and mechanical work performed when a stimulated muscle was allowed to shorten against varying loads. He then compared the energy liberated to isometric contractions and observed that a muscle liberated additional energy when allowed to shorten (Katz, 1978).

## Nikolai Bernstein - CNS Control of Posture and Movement

Nikolai Bernstein (1896 - 1966) is credited with demonstrating the underlying control mechanism that the central nervous system uses for both postural control and movement (Stuart, 2005). Bernstein advanced our understanding of the neural control of locomotion by studying the combined role of descending commands, sensory feedback and spinal pattern generators on the control of gait and posture.

## Andrew Huxley and Hugh Huxley - Sliding Filament Theory

The sliding filament theory of muscle contraction was developed independently by two scientists of the same last name Huxley, Andrew Fielding Huxley (1917 - 2012), and Hugh Esmor Huxley (1924 - 2013). According to the sliding filament theory of muscle contraction, muscle shortens by increasing the overlap between the protein filaments actin and myosin. Using mechanical energy obtained from the hydrolysis of ATP, myosin crossbridges form attachments with actin to pull actin towards the center of the sarcomere. The independent work of Andrew F. Huxley and Hugh E. Huxley describing the structure of actin and myosin was published in the journal **Nature** in 1954. Andrew Huxley used x-ray diffraction to image the structure of actin and myosin during isometric and isotonic contractions (Huxley, et al., 1954). Using an electron microscope Hugh Huxley provided the first images of the actin and myosin filaments (Huxley, et al., 1954). A few years later in 1957 Hugh Huxley provided proof of the existence of the crossbridges attaching myosin to actin (Huxley, 1957). Finally, in 1971, Huxley and Simmons published the sliding filament theory and explained how the myosin crossbridges generate force in muscle (Huxley, et al., 1971).

## PROFESSIONAL ORGANIZATIONS IN BIOMECHANICS

There are many professional organizations dedicated to the study of biomechanics. Below is a partial list.

- American Society of Biomechanics (ASB)
- International Society of Electrophysiological Kinesiology (ISEK)
- International Society of Biomechanics (ISB)
- International Society of Biomechanics in Sports (ISBS)
- International Society for Posture and Gait Research (ISPGR)

## Professional Biomechanics Journals

There are many professional journals dedicated to biomechanical topics. Below is a partial list.

- Applied Bionics and Biomechanics
- Bone
- Clinical Biomechanics
- Footwear Science
- Gait and Posture
- Human Movement Science
- Journal of Applied Physiology
- Journal of Biomechanical Engineering
- Journal of Biomechanics
- Journal of Electromyography & Kinesiology
- Journal of Experimental Biology
- Journal of Applied Biomechanics
- Sports Biomechanics

## Biomechanics Instrumentation

Biomechanists use a variety of instrumentation to analyze motion, such as motion capture systems, force platforms, balance systems, electromyography systems, and qualitative sport analysis systems.

In the 1980s we used high-speed 16 mm cameras to capture motion. The cameras were capable of filming up to 500 frames per second. A biomechanical analysis was done by first filming the motion to be analyzed. Then the film was sent off for processing, which took about two weeks. The processed film was analyzed a frame at a time using a projector and an x-y digitizer interfaced to a mainframe computer or a personal computer. The analysis was completed using a 16 mm projector and a digitizer to obtain x-y coordinates of a body landmark. Each frame of film was displayed on a screen and a cursor similar to a mouse was used to get x, y coordinates. It would take approximately one hour to enter 2D x, y coordinates for 3 seconds of data. The user would digitize body landmarks (toe, ankle, knee, hip, shoulder, elbow, wrist) for an individual frame of film and then advance the projector to the next frame of film.

Today we use high-speed digital cameras to track x, y motion of a video marker. The motion is typically filmed with 6-12 cameras at rates of 60-1000 frames per second. The 2D marker coordinates are transformed into 3D (x, y, z) coordinates with the requirement that each marker is seen by two or more cameras.

Each section below describes the general categories of instrumentation used today and some examples of the companies that make them.

## Motion Capture Systems

Motion capture systems are used to record the motions of a single body part or the entire body. These systems use high speed cameras to capture motion and compute 3D coordinates of body landmarks. From the external body landmarks computer software is used to compute anatomically correct joint coordinate systems, which are used to compute kinematics and kinetics of motion. The following is a partial list of vendors for motion capture system: Motion Analysis Corporation, Qualisys, and Vicon Motion Systems.

## Electromyography Systems

Electromyography Systems (EMG) – The electrical signals that arise from muscular contractions are recorded using electromyography (EMG) amplifiers and specialized software which is then used to analyze the electrical signals. Two examples of vendors of EMG systems are Delsys and Noraxon.

## Force Plate Systems

Force Plate Systems – The forces associated with running, jumping, and balancing are collected and analyzed using force platform systems. Examples of force plate system vendors include AMTI, Bertec, and Kistler.

## Balance and Postural Control Systems

Balance and Postural Control – Balance and postural control are collected and analyzed using computerized dynamic posturography (CDP). Examples of balance and postural control system vendors include AMTI, Bertec, Biodex, and NeuroCom.

## Coaching and Physical Education Teaching Tools

Coaching and Physical Education Teaching Tools – Coaches and physical education professionals use video cameras with the specialized software Dartfish to quantify motion, compare performance, and provide the learner with mechanically correct feedback. Coaches and teachers also use a free program Kinovea, available from www.kinovea.org for video analysis. The Kinovea software package can be used to scale video images and compute kinematic and kinetic variables such and maximum velocity and peak power production during a parallel squat. Using an iPhone or Android based smart phone, coaches and teachers can video a motion and then analyze it for free with the Kinovea software.

### REFERENCES

Braune W, Fischer O. The human gait. trans. P Maquet & R Furlong, (Original manuscript published 1895 - 1904), Berlin: Springer-Verlag, 1987.

Hawking, Stephen. On the Shoulders of Giants. Philadelphia, PA: Running Press, 2002.

Haas, Robert B. Muybridge: Man in Motion. Berkeley: University of California Press, 1976.

Hill AV. The heat produced in contracture and muscular tone. J Physiol. 40: 389-403, 1910.

Hill AV. The heat of shortening and the dynamic constants of muscle. Proceedings of the Royal

Society of London, series B - Biolgical Sciences. 117: 136-195, 1938.

Hill AV, Hartree W. The four phases of heat-production of muscle. J Physiol. 54: 84-128, 1920.

Huxley AF, Niedergerke R. Structural changes in muscle during contraction. Nature. 173: 974-973, 1954.

Huxley AF, Simmons RM. Proposed mechanism of force generation in striated muscle. Nature. 233: 533-538, 1971.

Huxley HE. The double array of filaments in cross-striated muscle. The Journal of Biophysical and Biochemical Cytology. 3: 631-648, 1957.

Huxley HE, Hanson JM. Changes in the cross-striations of muscle during contraction and stretch and their structural interpretation. Nature. 173: 973-976, 1954.

Katz B. Archibald Vivian Hill. 26 September 1886-3 June 1977. Biographical Memoirs of Fellows of the Royal Society. 24: 71-149, 1978.

Martin, R. Bruce. A genealogy of biomechanics. 23rd Annual Conference of the American Society of Biomechanics. University of Pittsburgh, Pittsburgh PA: American Society of Biomechanics, 1999.

Molnar Z, Brown RE. Insights into the life and work of sir Charles Sherrington. Nat Rev Neurosci. 11: 429-436, 2010.

Newton, Isaac. The Principia: Mathematical Principles of Natural Philosophy. trans. I Bernard Cohen, Anne Whitman (Berkley, CA: University of California Press, 1999), 1687.

Pope MH. Giovanni Alfonso Borelli--the father of biomechanics. Spine. 30: 2350-2355, 2005.

Stuart DG. Integration of posture and movement: Contributions of Sherrington, Hess, and Bernstein. Human Movement Science. 24: 621-643, 2005.

Van der Kloot W. Mirrors and smoke: A. V. Hill, his brigands, and the science of anti-aircraft gunnery in world war I. Notes and Records of the Royal Society. 65: 393-410, 2011.

# Chapter 2

## VECTORS AND SCALARS

### STUDENT LEARNING OUTCOMES

**After reading Chapter 2, you should be able to**

- Define scalars and vectors.
- Compute x and y components of a vector.
- Compute the magnitude and direction of a vector.
- Graphically add two or more vectors.
- Mathematically add two or more vectors.

### DEFINITION OF SCALARS AND VECTORS

Scalars are quantities that can be completely described by magnitude alone. Distance, mass, work ,and volume are all examples of scalars.

Unlike scalars, to completely describe a vector, both direction and magnitude are needed. Examples of vectors include force, torque, momentum, position, velocity, and acceleration. Figure 2.1 shows a force vector with a magnitude of 100 N and a direction of 140 degrees. The convention in biomechanics is to define angles in the counterclockwise (CCW) direction from a right horizontal. The Greek symbol theta ($\theta$) is used to represent an angle.

The force vector shown in Figure 2.1 has both a horizontal and vertical component. Figure 2.2 illustrates the graphical method used to determine the x and y components of a vector when given the magnitude and direction of the vector. The x component of the vector is determined by drawing a vertical line down from the tip of the force vector to the x axis (see the dashed vertical line shown in Figure 2.2). The y component of the vector is determined by a horizontal line from the tip of the force vector to the y axis (see the dashed horizontal line shown in Figure 2.2). As shown in Figure 2.2, the x component of the force vector points to the left and the y component of the force vector points upward.

When drawing a vector, the length of the arrow describes the magnitude of the vector and the angle to the right horizontal defines the direction of a vector. Vectors A and B in Figure 2.3 both have a magnitude of 10 m/s and therefore they are each drawn with the same

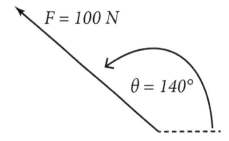

**Figure 2.1** Example of a vector. The vector *F* has a magnitude of 100 N and a direction of 140 degrees to the right horizontal.

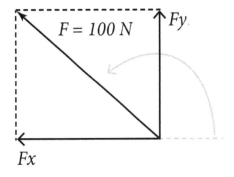

**Figure 2.2** X and Y components of a vector.

length arrow. Vector C in Figure 2.3 has a magnitude of 5 m/s and vector D has a magnitude of 15 m/s; notice that vector C is one-half as long as vectors A and B and that vector D is three times longer than vector C. The direction of vectors A – D in Figure 2.3 are each defined by the angle of the arrow to the right horizontal.

## VECTOR EQUATIONS

Figure 2.4 shows the vector equations used to get the x and y components or the direction and magnitude of a vector. We will use the trigonometric functions cosine (cos), sine (sin), and tangent (tan) for vector equtions. To determine the x and y components of a vector from its magnitude and direction, use the following equations:

$$x = r\cos\theta$$
$$y = r\sin\theta$$

where r is the magnitude, θ is the direction, x is the x component of the vector and y is the y component of the vector. To determine the magnitude and direction of a vector from its x and y components, use the following equations:

$$r = \sqrt{x^2 + y^2}$$
$$\theta = \tan^{-1}\left(\frac{y}{x}\right)$$

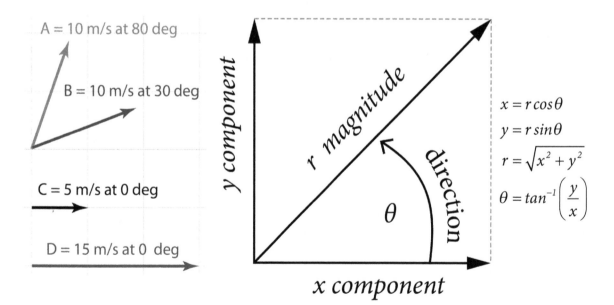

$$x = r\cos\theta$$
$$y = r\sin\theta$$
$$r = \sqrt{x^2 + y^2}$$
$$\theta = \tan^{-1}\left(\frac{y}{x}\right)$$

**Figure 2.3** The length of each arrow describes the vector magnitude. The vector direction is defined by the angle to the right horizontal.

**Figure 2.4** Vector equations.

where r is the magnitude of the vector, and theta ($\theta$) is the direction of the vector.

Before attempting the examples in this chapter *verify that your calculator is in "Degree Mode"*. All of the examples in this chapter assume that trigonomic functions (sin, cos, tan⁻¹) are calculated using degrees.

## DEFINING AN ANGLE CCW FROM A RIGHT HORIZONTAL

In biomechanics, angles are typically defined in the counterclockwise direction from a right horizontal. When using the tan⁻¹ function to determine the direction of a vector, your calculator will only give angles between −90° to +90° since the inverse tangent is restricted to the range of −90° to +90°. Figure 2.5 explains the output of the inverse tangent function using a vector with an x component of ±5.00 and a y component of ±8.66. Figure 2.6 depicts the transformations needed to convert angles that are not in the first quadrant so that they are defined in the CCW direction from the right horizontal. If the vector is in the second or third

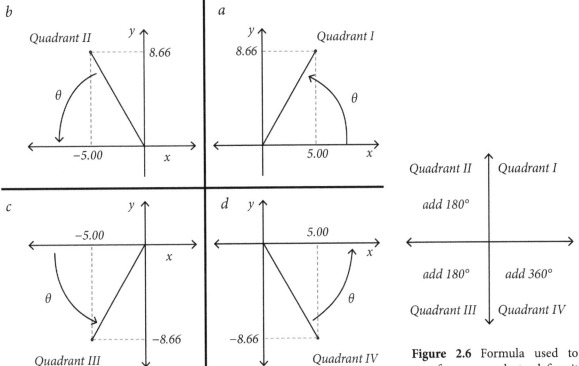

**Figure 2.5** Explanation of the output angle from tan-¹ function depending upon the quadrant the vector is located in.

**Figure 2.6** Formula used to transform an angle to define it from a right horizontal in the CCW direction.

quadrant we will add 180° (see Figure 2.6). For vectors in the fourth quadrant, add 360°.

$$\theta = tan^{-1}\left(\frac{y}{x}\right)$$

$$\theta = tan^{-1}\left(\frac{8.66}{5.00}\right)$$

$$\theta = 60°$$

In Figure 2.5a, the vector has components of (+5.00, +8.66), and using tan⁻¹ to compute the angle gives the following:

Since this angle is already defined in the CCW direction from the right horizontal, no trans-

$$\theta = tan^{-1}\left(\frac{y}{x}\right)$$

$$\theta = tan^{-1}\left(\frac{8.66}{-5.00}\right)$$

$$\theta = -60°$$

formation is needed. In Figure 2.5b, the vector has components of (−5.00, 8.66). Using tan⁻¹ to compute the angle gives the following:

$$\theta = -60° + 180°$$

$$\theta = +120°$$

The vector in Figure 2.5b is in the second quadrant. As shown in Figure 2.6, we will add 180° to transform the angle so that it is defined from right horizontal in the CCW direction as follows:

$$\theta = tan^{-1}\left(\frac{y}{x}\right)$$

$$\theta = tan^{-1}\left(\frac{-8.66}{-5.00}\right)$$

$$\theta = 60°$$

In Figure 2.5c, the vector has components of (−5.00, −8.66). Using tan⁻¹ to compute the angle gives the following:

$$\theta = 60° + 180°$$

$$\theta = 240°$$

In this case, the vector is in the third quadrant and we will add 180° (see Figure 2.6) to transform the angle so that it is defined from right horizontal in the CCW direction as follows:

$$\theta = tan^{-1}\left(\frac{y}{x}\right)$$

$$\theta = tan^{-1}\left(\frac{8.66}{-5.00}\right)$$

$$\theta = -60°$$

Finally, in Figure 2.5d, the vector has components of (+5.00, −8.66). Using tan⁻¹ to compute the angle gives the following:

$$\theta = -60° + 360°$$

$$\theta = +300°$$

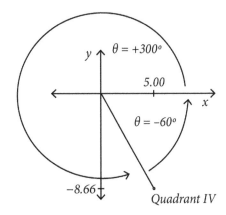

In this case, the vector is in the fourth quadrant. As shown in Figure 2.6, we will add 360° to transform the angle so that it is defined from right horizontal in the CCW direction as follows.

## COMPUTING X AND Y COMPONENTS OF A VECTOR

### X and Y Components Example 1

Compute the x and y components of a 100 N force acting at an angle of 140 degrees, see Figure 2.7. To get the x component of a vector, multiply the magnitude by the cosine of its direction as follows:

$$x = r(cos\theta)$$

$$Fx = F(cos\theta)$$

$$Fx = 100N(cos\,140)$$

$$Fx = -76.60N$$

$$y = r(sin\theta)$$

$$Fy = F(sin\theta)$$

$$Fy = 100N(sin\,140)$$

$$Fy = 64.28N$$

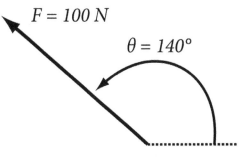

**Figure 2.7** Determine the x and y components of a vector with a magnitude of 100 N and a direction of 140°.

The y component of the force vector F is obtained by multiplying the magnitude (100 N) by the sine of its direction (140°) as follows:

Notice that the force vector F is in the second quadrant. Therefore it has a positive y coordinate ($Fy = 64.28$ N) and a negative x component ($Fx = -76.60$ N).

### X and Y Components Example 2

Compute the x and y components of a position vector $P$ with a magnitude of 4 m and a direction ($\theta$) acting at an angle of 300° (see Figure 2.8). To get the x component of a vector, multiply the magnitude by the cosine of its direction as follows.

$$x = r(\cos\theta)$$
$$Px = P(\cos\theta)$$
$$Px = 4m(\cos 300)$$
$$Px = 2.00m$$

The y component of the position vector $P$ is obtained by multiplying the magnitude (4 m) by the sin of its direction (300°) as follows.

$$y = r(\sin\theta)$$
$$Py = P(\sin\theta)$$
$$Py = 4m(\sin 300)$$
$$Py = -3.46m$$

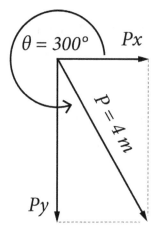

**Figure 2.8** Determine the x and y components of a vector with a magnitude of 4 m and a direction of 300°.

Notice that the position vector P is in the fourth quadrant. Therefore it has a positive x coordinate ($Px = 2.00$ m) and a negative y component ($Py = -3.46$ m).

### Computing the Magnitude and Direction of a Vector from X and Y Components

To compute the magnitude and direction of a vector from its x and y components use the Pythagorean theorem to get the magnitude and the inverse tangent function to get the direction.

### Magnitude & Direction Example 1

Determine the magnitude ($V$) and direction ($\theta$) of a velocity vector with an x component $Vx$ of 4 m/s and a y component $Vy$ of 6 m/s. See Figure 2.9 for an illustration of the vector components. The magnitude of the vector is obtained as follows:

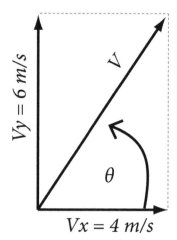

**Figure 2.9** Determine the direction $\theta$ and magnitude $V$ of a vector with components $Vx$ of 4 m/s and $Vy$ of 6 m/s,

$$r = \sqrt{x^2 + y^2}$$
$$V = \sqrt{Vx^2 + Vy^2}$$
$$V = \sqrt{(4.0)^2 + (6.0)^2}$$
$$V = 7.21 m/s$$

The direction ($\theta$) of the velocity vector is computed as follows.

$$\theta = tan^{-1}\left(\frac{y}{x}\right)$$
$$\theta = tan^{-1}\left(\frac{Vy}{Vx}\right)$$
$$\theta = tan^{-1}\left(\frac{6m/s}{4m/s}\right)$$
$$\theta = 51.34°$$

The velocity vector (V) is located in the first quadrant since the x and y components of the vector are both positive. As shown in Figure 2.6, no change is required to convert the vector so that it is measured counterclockwise from the right horizontal.

## Magnitude & Direction Example 2

Determine the magnitude (A) and direction ($\theta$) of an acceleration vector with an x component $Ax$ of −8 m/s² and a y component $Ay$ of −10 m/s². See Figure 2.10 for an illustration of the vector components. The magnitude of the vector is obtained as follows:

$$r = \sqrt{x^2 + y^2}$$
$$A = \sqrt{Ax^2 + Ay^2}$$
$$A = \sqrt{(-8.0)^2 + (-10.0)^2}$$
$$A = 12.81 m/s^2$$

The direction ($\theta$) of the acceleration vector (A) is computed as follows:

$$\theta = tan^{-1}\left(\frac{y}{x}\right)$$
$$\theta = tan^{-1}\left(\frac{Ay}{Ax}\right)$$
$$\theta = tan^{-1}\left(\frac{-10m/s^2}{-8m/s^2}\right)$$
$$\theta = 51.34°$$
$$\theta = 51.34° + 180°$$
$$\theta = 231.34°$$

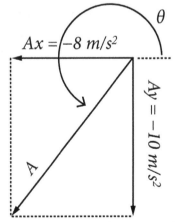

**Figure 2.10** Compute the magnitude (A) and the direction ($\theta$) of an acceleration vector with an x component $Ax$ = −8 m/s² and a y component $Ay$ = −10 m/s².

The acceleration vector (A) is located in the third quadrant since the x and y components of the vector are both negative. As shown in Figure 2.4, add 180° to the inverse tangent function output of 51.34° to convert the vector so that it measured counterclockwise from the right horizontal. The acceleration vector has a magnitude of 12.81 m/s$^2$ and a direction of 231.34°.

## GRAPHICALLY ADDING TWO OR MORE VECTORS

To graphically add two or more vectors, use the "tip to tail" method. Draw the first vector, place the tail of the second vector on the tip of vector 1, and then place the third vector tail on the tip of vector 2. Finally, draw a vector from where you started (tail of vector 1) to where you finished (tip of vector 3). See the Figure 2.11 for an example of adding three vectors. The dashed arrow is the resultant vector when adding A + B + C. Adding the vectors in a different order such as C + B + A or B + A + C will still give the same result.

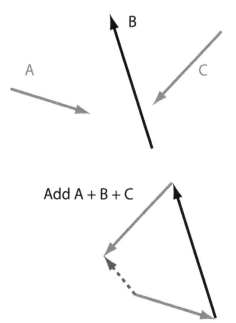

**Figure 2.11** Using the "tip to tail" method to graphically add vectors A, B, and C.

Before introducing how to mathematically add vectors, visually inspect the vectors and the result shown in Figure 2.11. Notice that the x component of C is greater and opposite of the x component of A. Therefore the addition of $Ax + Cx$ will give a small negative number which will then be added to the small negative x component of vector B. The combination of $Ax + Bx + Cx$ produces in the x component of the resultant vector (arrow with dashed line). Similarly, the addition of $Cy$ with $Ay$ results in a small negative y component which is then added to the relatively large $By$ component. Thus the addition of $Ay + By + Cy$ yields the y component of the resultant vector (arrow with dashed line).

## MATHEMATICALLY ADDING TWO OR MORE VECTORS

Figure 2.12 shows the mathematical solution for the graphical solution that was presented in Figure 2.11. To mathematically add two or more vectors, follow these steps:

1.  Get the x and y components of each vector using the sine and cosine.

$$x = r \cos \theta$$

$$y = r \sin \theta$$

2.  Add the x components together to get a resultant x component ($Rx$).

$$Rx = Ax + Bx + Cx + ...$$

3.  Add the y components together to get a resultant y component ($Ry$).

$$Ry = Ay + By + Cy + ...$$

4.  Get the magnitude of the resultant vector using the Pythagorean theorem.

$$r = \sqrt{x^2 + y^2}$$

$$r = \sqrt{Rx^2 + Ry^2}$$

5.  Get the direction of the resultant vector using the inverse tangent function.

$$\theta = \tan^{-1}\left(\frac{y}{x}\right)$$

$$\theta = \tan^{-1}\left(\frac{Ry}{Rx}\right)$$

6.  Convert the angle to 0–360° measured CCW from the right horizontal depending upon the quadrant of the resultant vector.

Figure 2.12 illustrates how to mathematically add three vectors: vector A with a magnitude of 108 m at 343°, vector B with a magnitude of 190 m at 108°, and vector C with a magnitude of 132 m at 226°. The resultant vector has a magnitude of 71.8 m with a direction of 131.02°.

## SUMMARY

This chapter explains the difference between a vector and a scalar. The vector equations to compute x and y components of a vector and the magnitude and direction of a vector were described. Finally, the addition of two or more vectors was demonstrated both graphically and mathematically.

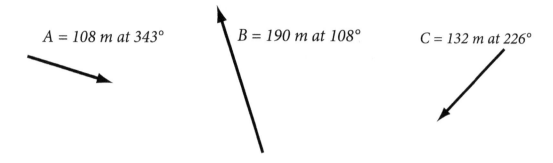

$A = 108\ m\ at\ 343°$       $B = 190\ m\ at\ 108°$       $C = 132\ m\ at\ 226°$

$Ax = r(cos\theta)$       $Ay = r(sin\theta)$       $Bx = r(cos\theta)$       $By = r(sin\theta)$       $Cx = r(cos\theta)$       $Cy = r(sin\theta)$

$Ax = 108(cos\,343°)$  $Ay = 108(sin\,343°)$  $Bx = 190(cos\,108°)$  $By = 190(sin\,108°)$  $Cx = 132(cos\,26°)$  $Cy = 132(sin\,226°)$

$Ax = 103.28m$       $Ay = -31.58m$       $Bx = -58.71m$       $By = 180.70m$       $Cx = -91.69m$       $Cy = -94.95m$

$Rx = Ax + Bx + Cx$                    $Ry = Ay + By + Cy$

$Rx = 103.28 - 58.71 - 91.69$           $Ry = -31.58 + 180.70 - 94.95$

$Rx = -47.13m$                         $Ry = 54.17m$

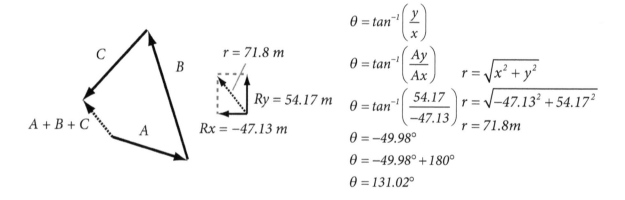

$r = 71.8\ m$

$Ry = 54.17\ m$

$Rx = -47.13\ m$

$\theta = tan^{-1}\left(\dfrac{y}{x}\right)$

$\theta = tan^{-1}\left(\dfrac{Ay}{Ax}\right)$       $r = \sqrt{x^2 + y^2}$

$\theta = tan^{-1}\left(\dfrac{54.17}{-47.13}\right)$   $r = \sqrt{-47.13^2 + 54.17^2}$

$\theta = -49.98°$                       $r = 71.8m$

$\theta = -49.98° + 180°$

$\theta = 131.02°$

**Figure 2.12** Result of mathematical addition of three vectors A, B, C.

## REVIEW QUESTIONS

Note: The answers are given in square brackets [].
1. Define a scalar and give an example of a scalar.
2. Compute the x, y components of an position vector with a magnitude of 4 m at an angle of 300 degrees.
3. Define a vector and give an example of a vector.
4. Draw a picture to show the result of graphically adding 3 m/s at 30 degrees with 6 m/s at 180 degrees.
5. A force of 100 N is applied at an angle of 60 degrees. Draw the vector and its x, y components.

6. Compute the x, y components of a force vector F = 400 N at 210 degrees.
7. A soccer ball is accelerated with a horizontal component Ax = 7 m/s$^2$ and Ay = 14 m/s$^2$, draw the x, y components and the acceleration vector direction and magnitude.
8. Compute the direction and magnitude of an acceleration vector with components Ax = 7 m/s$^2$ and Ay = −6 m/s$^2$.
9. Compute the x and y components of each vector. Draw the vector and the x, y components.
   a. 200 N at 30 deg  [x = 173.21 N, y = 100 N]
   b. 4 m/s at 220 deg  [x = −3.06 m/s, y = −2.57 m/s]
10. Compute the direction and magnitude of an acceleration vector with components Vx = −11 m/s and Vy = 8 m/s.
11. Compute the magnitude and direction of each vector.  If the vector is not in the first quadrant convert the angle so that it is defined counterclockwise from the right horizontal.
   a. Fx = 200 N, Fy = 1400 N  [magnitude = 1414.21 N, direction = 81.87°]
   b. Ax = −2 m/s$^2$, Ay = −4 m/s$^2$  [magnitude = 4.47 m/s$^2$, direction = 243.43°]
   c. Vx = −3 m/s, Vy = 6 m/s  [magnitude = 6.71 m/s, direction = 116.57°]
12. Compute the direction and magnitude of the addition of A = 100 N at 40 degrees with B = 180 N at 330 degrees.
13. Add the vector pairs shown below.  Give a graphical picture of the vector addition and compute the magnitude and direction of the resultant vector.  If the resultant vector is not in the first quadrant convert the angle so that it is defined counterclockwise from the right horizontal.
   a. Vector A = 300 N at 50 degrees, Vector B = 100 N at 25 degrees.  [magnitude = 392.91 N, direction = 43.83°]
   b. Vector A = 10 m at 310 degrees, Vector B = 3 m at 140 degrees.  [magnitude = 7.06 m, direction = 305.77°]

## SUGGESTED SUPPLEMENTAL RESOURCES

Cutnell, J. D. and K. W. Johnson (2008). Physics. Hoboken, John Wiley & Sons.

Enoka, R. M. (2008). Neuromechanics of Human Movement. Champaign, Human Kinetics.

Griffiths, I. W. (2006). Principles of Biomechanics & Motion Analysis. Philadelphia, Lippincott Williams & Wilkins.

Hall, S. J. (2007). Basic Biomechanics. Boston, McGraw Hill.

Hamill, J. and K. M. Knutzen (2009). Biomechanical Basis of Human Movement. Philadelphia, Lippincott Williams & Wilkins.

Hibbeler, R. C. (2010). Engineering Mechanics: Dynamics. Upper Saddle River, Prentice Hall.

Hibbeler, R. C. (2010). Engineering Mechanics: Statics. Upper Saddle River, Prentice Hall.

McGinnis, P. M. (2013). Biomechanics of Sport and Exercise. Champaign, Human Kinetics.

# Chapter 3

## LINEAR KINEMATICS

### STUDENT LEARNING OUTCOMES

**After reading Chapter 3, you should be able to:**

- Define kinematics, position, velocity, and acceleration.
- Analyze the slope of a position curve to estimate velocity.
- Analyze the slope of a velocity curve to estimate acceleration.
- Analyze the area of an acceleration curve to determine the change in velocity.
- Analyze the area of a velocity curve to determine the change in position.
- Solve linear kinematics problems.

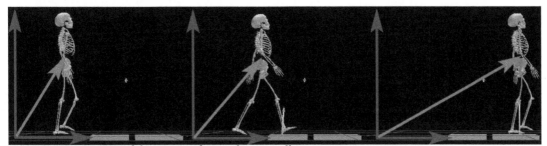

**Figure 3.1** Position vector of the center of mass during walking.

## POSITION

Kinematics is the description of motion using position, velocity, and acceleration. This lesson will focus on linear kinematics, or motion in a straight line. The position is usually measured relative to a starting position or an x-y axis. Position is a vector; therefore it has both direction and magnitude. Figure 3.1 shows the position vector of the center of mass for a subject walking. The center of mass position increases as the subject gets further and further away from the x, y origin.

Kinematic data are usually captured in the laboratory by filming a subject doing the motion of interest. x, y coordinates of body landmarks are obtained from the video data. The x, y coordinates represent the position of the body with respect to time.

## VELOCITY

Velocity is defined as the rate of change in position. Velocity is computed using:

$$Velocity = \frac{\Delta Position}{\Delta Time}$$

or

$$V = \frac{P_f - P_i}{\Delta t}$$

where $P_f$ is final position, and $P_i$ is initial position.

## RELATIONSHIP BETWEEN SLOPE AND VELOCITY

Velocity can also be described as the slope of the position–time graph. The instantaneous velocity at any point in time can be computed by the slope of a tangent line from a position–time graph (see Figure 3.1). The tangent line at a point is defined as the line that intersects the curve at only one point. The position–time graph shown in Figure 3.2 shows the relationship between slope and velocity. The slope (velocity) is 8 m/s at a time of 4 seconds and 16 m/s at a t = 8 seconds. In Figure 3.2, the slope of a tangent line intersecting the position–time curve at t = 4 seconds gives a velocity of 8 m/s. The slope at t = 8 seconds is greater than the slope at t = 4 s. The resulting velocity at t = 8 s is 16 m/s.

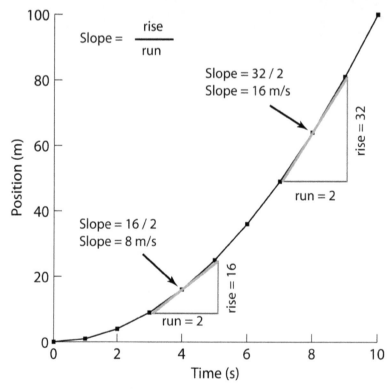

**Figure 3.2** Velocity is defined as the slope of the position-time curve. The instantaneous velocity is the slope of a tangent line at any given point where the tangent line intersects the curve.

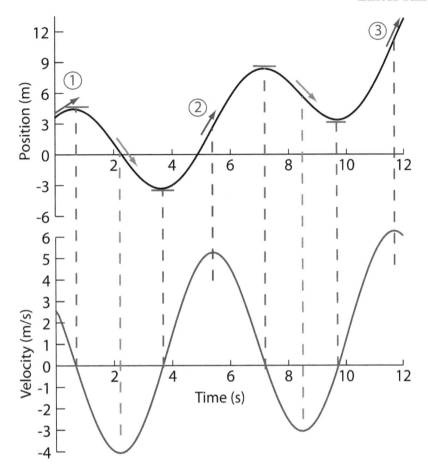

**Figure 3.3** The relationship between slope of a position – time curve and the resulting velocity. When the slope is horizontal the corresponding velocity is zero. When the slope is positive (up arrows) the corresponding velocity is positive. When the slope is negative (down arrows) the corresponding velocity is negative. The circled numbers 1, 2, and 3 illustrate that when the slope is greater the resulting velocity will have a higher magnitude.

## COMPUTING A DERIVATIVE

The process of evaluating the slope to get the rate of change is called taking a derivative. Figure 3.3 depicts the relationship between the slope of a position–time curve and the corresponding velocity. The basic rules for estimating the derivative from a position–time graph are:
1. If the slope is horizontal, the velocity is 0.
2. If the slope is positive (arrow pointing upward), the velocity is positive.
3. If the slope is negative (arrow pointing downward), the velocity is negative.

When the slope is 0, shown by the horizontal lines in the graph in Figure 3.3, the velocity is 0. Follow the dashed lines from the horizontal slope locations on the position–time curve to the velocity–time curve to verify that the velocity is zero whenever the slope of the position–time graph is horizontal.

When the slope is positive (arrow pointing upward), the velocity is positive. Follow the dashed lines from the upward sloped arrows on the position–time curve to the velocity–time curve to verify that the velocity is positive whenever the slope of the position – time graph is positive.

The circled numbers 1, 2, and 3 on the position–time graph illustrate points with three different slopes. In each case, the slope is positive and therefore the corresponding velocity is positive. At point 1, the slope is less than the slope at point 2. The corresponding velocities for these points show the same relationship, the velocity for point 1 is less than the velocity for point 2. Finally, the slope at point 3 is the greatest and the corresponding velocity at point 3 is also greater than the velocities at points 1 and 2.

When the slope is negative (arrow pointing downward), the velocity is negative. Follow the dashed lines from the downward arrows on position–time curve to the velocity–time curve to verify that the velocity is negative whenever the slope of the position–time graph is negative.

### ACCELERATION

Acceleration is the rate of change in velocity, or the slope of the velocity–time graph. Acceleration is defined with the following formulas:

$$Acceleration = \frac{\Delta Velocity}{\Delta Time}$$

or

$$A = \frac{V_f - V_i}{\Delta t}$$

where $V_f$ is final velocity, and $V_i$ is initial velocity.

### RELATIONSHIP BETWEEN SLOPE OF VELOCITY GRAPH AND ACCELERATION

The slope of a velocity–time curve at any point gives the corresponding acceleration for that point. In other words, computing the derivative (slope) of velocity gives the acceleration. Figure 3.4 shows the relationship between the slope of a velocity–time curve and the corresponding acceleration–time curve. The rules for evaluating the slope of a velocity–time curve to determine acceleration are exactly the same as those for evaluating the slope of a position–time curve to determine velocity.

The basic rules for estimating the derivative of the velocity are:
1. If the slope is horizontal (horizontal line), the acceleration is 0.
2. If the slope is positive (arrow pointing upward), the acceleration is positive.
3. If the slope is negative (arrow pointing downward), the acceleration is negative.

When the slope is horizontal, shown by the horizontal lines on the velocity graph in Figure 3.4, the acceleration is 0. Follow the dashed lines from the horizontal slope locations on the velocity - time curve to the acceleration - time curve to verify that the acceleration is zero whenever the slope of the velocity–time graph is horizontal.

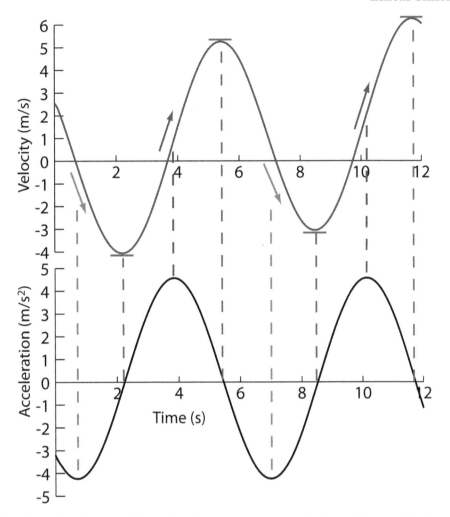

**Figure 3.4** The relationship between slope of a velocity–time curve and the resulting acceleration. When the slope is horizontal (horizontal lines) the corresponding acceleration is zero. When the slope is positive (upward arrows) the corresponding acceleration is positive. When the slope is negative (downward arrows) the corresponding acceleration is negative.

When the slope is positive (arrow pointing upward), the acceleration is positive. Follow the dashed lines from the upward sloped arrow locations on the velocity–time curve to the acceleration–time curve to verify that the acceleration is positive whenever the slope of the velocity–time graph is positive.

When the slope is negative (arrow pointing downward), the acceleration is negative. Follow the dashed lines from the downward arrow locations on the velocity–time curve to the acceleration–time curve to verify that the acceleration is negative whenever the slope of the velocity–time graph is negative.

As explained above, the mathematical description of taking a derivative involves evaluating the slope of a function versus time (position–time, or velocity–time) to determine the rate of change in the variable of interest (position or velocity). Briefly examine the graph above to review the relationship between acceleration and velocity. Notice that in every interval where the acceleration is positive, the corresponding velocity increases over that interval and for every interval where the acceleration is negative, the corresponding velocity decreases over that interval.

## INTEGRATION OF ACCELERATION TO OBTAIN CHANGE IN VELOCITY

When given acceleration–time curve it is possible to work backward to obtain the change in velocity. Integration is the mathematical process that is used to get the change in velocity from a acceleration–time curve or the change in position from a velocity–time curve. Integration of an acceleration–time curve is accomplished by getting the area between the acceleration curve and the y = 0 axis over some time interval.

The mathematical formula used to integrate some function with respect to time usually written using the integral sign ( ∫ ). To obtain the change in velocity from an acceleration–time curve, the integration would be written as follows.

$$V = \int_{t_0}^{t_1} A \, dt$$

Where V is the velocity, A is the acceleration and ∫ Is the integral of acceleration over the interval from t0 to t1. There are several methods that can be used to get the area of a function over time. One such method is to use the formula for the area of a rectangle.

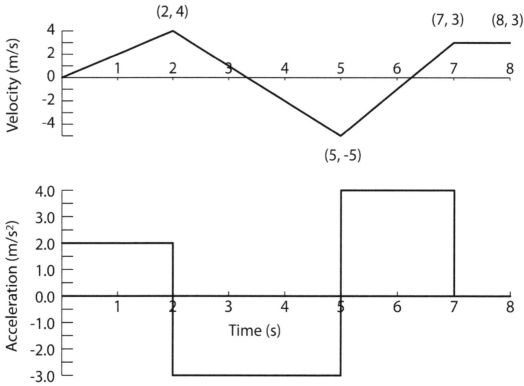

**Figure 3.5** Integration of acceleration with respect to time gives the change in velocity over the corresponding time interval.

Integration of the acceleration curve shown in Figure 3.5 with respect to time, using the formula for the area of a rectangle (height × width), will give the ***change in velocity*** over each time interval. The acceleration is +2 m/s² over the time interval from t = 0 to t = 2.

$$V = \int_0^2 A\,dt$$

The change in velocity over the time interval from 0 s to 2 s is equal to the area of the acceleration over the same time interval.

$$Area = Height \times Width$$
$$\Delta Velocity_{t=0}^{t=2} = 2m/s^2 \times 2s$$
$$\Delta Velocity_{t=0}^{t=2} = 4m/s$$

As a result, if the velocity was initially at 0 m/s for t = 0, the velocity would be 4 m/s at a time of 2 s. The x, y coordinates for each velocity–time point are shown in parentheses at each time point in Figure 3.5.

For the next interval, from t = 2 to t = 5, the integral of acceleration is:

$$V = \int_2^5 A\,dt$$

The change in velocity over the time interval from 2 s to 5 s is equal to the area of the acceleration over the same time interval.

$$Area = Height \times Width$$
$$\Delta Velocity_{t=2}^{t=5} = -3m/s^2 \times 3s$$
$$\Delta Velocity_{t=2}^{t=5} = -9m/s$$

As a result, if the velocity must decrease by −9 m/s over the interval, the velocity is −5 m/s at a time of 5 seconds. Over the next interval from t = 5 s to t = 7 s, the acceleration is +4 m/s². The change in velocity over this interval is:

$$V = \int_5^7 A\,dt$$

The change in velocity over the time interval from 5 s to 7 s is equal to the area of the acceleration over the same time interval.

$$Area = Height \times Width$$
$$\Delta Velocity_{t=5}^{t=7} = 4m/s^2 \times 2s$$
$$\Delta Velocity_{t=5}^{t=7} = 8m/s$$

As a result, if the velocity must decrease by 8 m/s over the interval, the velocity is +3 m/s at a time of 7 seconds. In the final interval from t = 7 s to t = 8 s, the acceleration is 0 m/s². Since the acceleration is 0 m/s² the velocity will not change over the interval from t = 7 s to t = 8 s. Therefore the velocity is constant at V = 3 m/s for the interval. The velocity graph shown in Figure 3.6 increases over all intervals where the acceleration is positive and the velocity decreases over all intervals where the acceleration is negative.

## INTEGRATION OF VELOCITY TO OBTAIN THE CHANGE IN POSITION

Integration of the velocity curve shown in Figure 3.6 with respect to time, using the formula for the area of a rectangle (height × width), will give the change in position over each time interval. The velocity is −3 m/s over the time interval from t = 0 to t = 2.

$$P = \int_0^2 V\, dt$$

The change in position over the time interval from 0 s to 2 s is equal to the area of the velocity over the same time interval.

$$Area = Height \times Width$$

$$\Delta Position_{t=0}^{t=2} = -3m/s \times 2s$$

$$\Delta Position_{t=0}^{t=2} = -6m$$

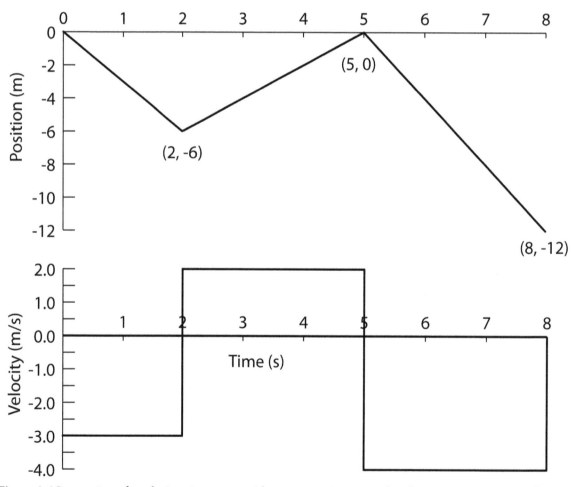

**Figure 3.6** Integration of a velocity–time curve with respect to time gives the change in position over the corresponding time interval.

As a result, if the position was initially at 0 m at t = 0 s, the position would be −6 m at a time of 2 seconds. The x, y coordinates for each position–time point are shown in parentheses at each time point in Figure 3.6. For the next interval, from t = 2 s to t = 5 s the integral of velocity is:

$$P = \int_{2}^{5} V \, dt$$

The change in position over the time interval from 2 s to 5 s is equal to the area of the velocity over the same time interval.

$$Area = Height \times Width$$

$$\Delta Position_{t=2}^{t=5} = 2m / s \times 3s$$

$$\Delta Position_{t=2}^{t=5} = 6m$$

As a result, if the position must increase by +6 m over the interval, the position is 0 m at a time of 5 s. Over the next interval, from t = 5 s to t = 8 s the velocity is −4 m/s. The change in position over this interval is:

$$P = \int_{5}^{7} V \, dt$$

The change in position over the time interval from 5 to 8 s is equal to the area of the velocity over the same time interval.

$$Area = Height \times Width$$

$$\Delta Position_{t=5}^{t=7} = -4m / s \times 3s$$

$$\Delta Position_{t=5}^{t=7} = -12m$$

As a result, if the position must decrease by −12 m over the interval, the position is −12 m at a time of 8 seconds. Finally, notice that the position decreases over any intervals where the velocity is negative and the position increases over any intervals where the velocity is positive.

## USING EXCEL TO COMPUTE VELOCITY AND ACCELERATION FROM POSITION

Figure 3.7 gives an example of computing velocity and acceleration from position–time data in Excel. This example will use the Excel file "**Get Vel & Accel.xls**" which is available on the course website. Velocity is defined as the rate of change in position. The formula to compute velocity is:

$$Velocity = \frac{\Delta Position}{\Delta Time}$$

or

$$V = \frac{P_f - P_i}{\Delta t}$$

where $P_f$ is final position, and $P_i$ is initial position. To compute velocity, enter the following equation in cell C3:

=(B3−B2)/0.1

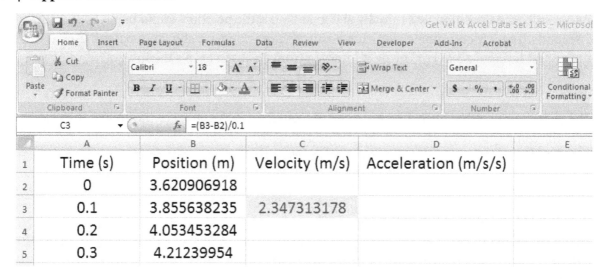

**Figure 3.7** Computing velocity from position-time data using Excel.

Then copy the formula down column C until row 122. Figure 3.7 shows how to compute the velocity that is associated with the change in position from t = 0.0 s to t = 0.1 s.

Acceleration is the rate of change in velocity, or the slope of the velocity–time graph. Acceleration is defined with the following formulas:

$$Acceleration = \frac{\Delta Velocity}{\Delta Time}$$

or

$$A = \frac{V_f - V_i}{\Delta t}$$

where $V_f$ is final velocity and $V_i$ is initial velocity. To compute acceleration, enter the following equation in cell D4:

=(C4−C3)/0.1

Then copy the equation down column D until row 122. Figure 3.8 displays the resulting acceleration at t = 0.2 s. When you have completed computing the velocity and acceleration, examine the x-y scatter graphs for each curve. You will see tabs on the bottom of the Excel worksheet that contain the position, velocity, and acceleration time graphs.

### INTEGRATION: COMPUTING VELOCITY FROM ACCELERATION IN EXCEL

Now we are going to work backwards to integrate the acceleration curve to compute the change in velocity. To obtain the change in velocity from an acceleration–time curve the integration would be written as follows:

| | A | B | C | D |
|---|---|---|---|---|
| 1 | Time (s) | Position (m) | Velocity (m/s) | Acceleration (m/s/s) |
| 2 | 0 | 3.620906918 | | |
| 3 | 0.1 | 3.855638235 | 2.347313178 | |
| 4 | 0.2 | 4.053453284 | 1.978150484 | -3.691626941 |
| 5 | 0.3 | 4.21239954 | 1.589462564 | |

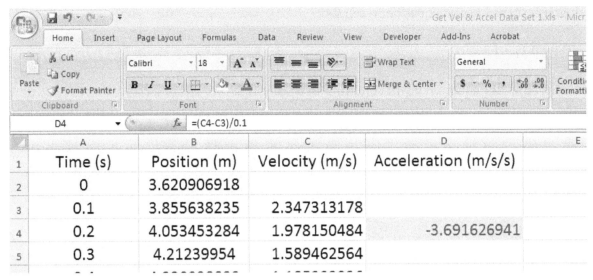

Figure 3.8 Computing acceleration from velocity-time data using Excel.

$$\Delta V = \int A\,dt$$

$$V_f - V_i = \int A \times dt$$

$$V_f = \int (A \times dt) + V_i$$

Figure 3.9 shows how to compute the change in velocity from acceleration–time data in Excel. Notice in Figure 3.9 that the initial velocity is given in cell E3. To implement the above equation, enter the initial velocity in cell E3 using the following equation: =C3

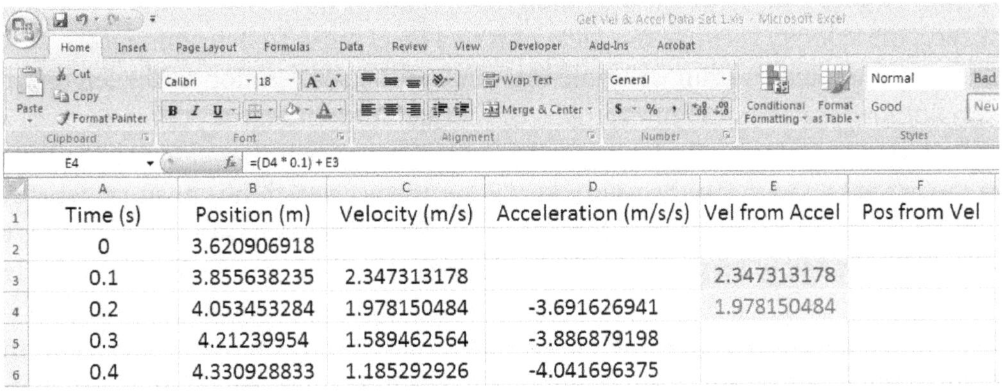

| | A | B | C | D | E | F |
|---|---|---|---|---|---|---|
| 1 | Time (s) | Position (m) | Velocity (m/s) | Acceleration (m/s/s) | Vel from Accel | Pos from Vel |
| 2 | 0 | 3.620906918 | | | | |
| 3 | 0.1 | 3.855638235 | 2.347313178 | | 2.347313178 | |
| 4 | 0.2 | 4.053453284 | 1.978150484 | -3.691626941 | 1.978150484 | |
| 5 | 0.3 | 4.21239954 | 1.589462564 | -3.886879198 | | |
| 6 | 0.4 | 4.330928833 | 1.185292926 | -4.041696375 | | |

**Figure 3.9** Integrating an acceleration-time curve to compute the change in velocity using Excel.

Then, enter the integral equation in cell E4.

=(D4 * 0.1) + E3

Finally, copy this equation down column E until row 122.

## INTEGRATION: COMPUTING POSITION FROM VELOCITY IN EXCEL

Now we are going to work backwards to integrate the velocity curve to compute the change in position. To obtain the change in position from a velocity–time curve the integration would be written as follows:

$$\Delta P = \int V \, dt$$
$$P_f - P_i = \int V \times dt$$
$$P_f = \int (V \times dt) + P_i$$

To implement the above equation enter the initial velocity in cell F2 using the following equation: =B2

Then, enter the integral equation in cell F3.

=(E3 * 0.1) + F2

Finally, copy this equation down column F until row 122. Now examine the values for position in columns B and F, they should be exactly the same. The values for the velocity in columns C and E should be the same as well.

## LOCAL MAX AND LOCAL MIN

It is possible to determine whether the velocity is at a local max or local min using the acceleration–time curve (see Figure 3.10). The dashed lines on the acceleration graph of Figure 3.10 indicate the locations where acceleration has a value of zero. When the acceleration has a value of zero, the velocity is either at a local min or a local max. To determine if the zero point of acceleration is a max or a min, examine the value of the acceleration to the left and right of the zero point for acceleration. For example, the acceleration in Figure 3.10 is zero at $t = 2.5$ s. The acceleration is negative to the left of $t = 2.5$, which means that the velocity has a downward slope to the left of $t = 2.5$ s. To the right of $t = 2.5$ s, the acceleration is positive, which means that the velocity has a positive slope to the right of $t = 2.5$ s. Therefore, since velocity has a negative slope to the left of $t = 2.5$ s and a positive slope to the right of $t = 2.5$ s, the velocity must be at a local min at $t = 2.5$ s. At $t = 5.35$ s the acceleration is at another zero point. At $t = 5.35$ s the acceleration is positive (velocity has upward slope) to the left of $t = 5.35$ s and negative (velocity has downward slope) to the right of $t = 5.35$ s, therefore velocity is at a local max.

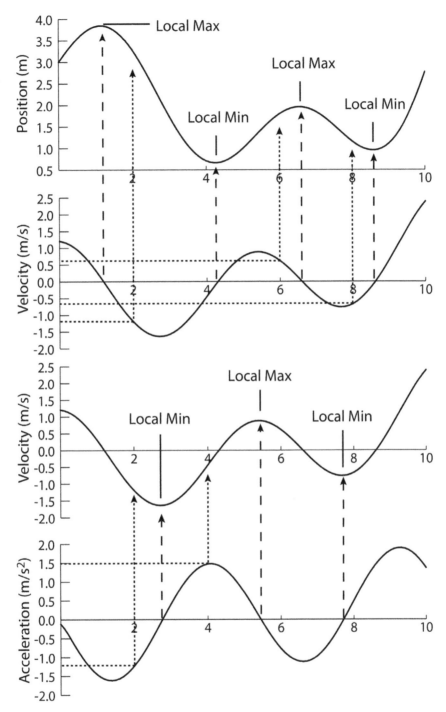

**Figure 3.10** Local max and local min for acceleration and velocity. In the acceleration curve at the bottom, a zero for acceleration is either a local max or min for velocity. If the acceleration is negative to the left of the zero point and positive to the right, then velocity is at a local min. If the acceleration is positive to the left of the zero point and negative to the right, then velocity is at a local max. The value of acceleration is the slope of velocity, at t=2 s the acceleration is -1.24 m/s², which is the slope of velocity at 2 s.

A velocity point with a value of zero is either a local max or local min for the position. The velocity and position time curves are shown in the top of Figure 3.10. The velocity is 0 m/s at t = 1.2 s. The velocity is positive to the left of t = 1.2 s and negative to the right of t = 1.2 s, so therefore, position is at a local max at t = 1.2 s. At t = 4.3 s the velocity is at another zero point. This time the velocity is negative to the left of t = 4.3 s and positive to the right of t = 4.3 s, so therefore, position must be at a local min at t = 4.3 s.

The dotted lines in Figure 3.10 show the relationship between the value of acceleration and the slope of velocity at a single time point. The value of the acceleration at any time point represents the slope of the velocity at that time point. For example, in Figure 3.10, the acceleration has a value of $-1.24$ m/s$^2$ at a time of t = 2 s, which means that the velocity at t = 2 s has a slope of $-1.24$ m/s$^2$. The acceleration has a value of $+1.47$ m/s$^2$ at a time of t = 4 s, which means that the velocity at t = 4 s has a slope of $+1.47$ m/s$^2$.

The relationship between the value of velocity and the slope is interpreted in the same manner. For example, the velocity has a value of $-1.20$ m/s at a time of t = 2 s, which means that the position at t = 2 s has a slope of $-1.24$ m/s. The velocity has a value of $+0.61$ m/s at a time of t = 6 s, which means that the position at t = 6 s has a slope of $+0.61$ m/s. Finally, the velocity has a value of $-0.67$ m/s at a time of t = 8 s, which means that the position at t = 8 s has a slope of $-0.67$ m/s.

## GRAPHICAL RELATIONSHIP BETWEEN ACCELERATION AND VELOCITY

The graph in Figure 3.11 shows the relationship between the area of an acceleration interval and the change in velocity for the same time interval. In the first interval, the area of acceleration is 2.85 m/s and the velocity decreases by 2.85 m/s over the same time interval. Knowledge of the area of acceleration does not tell you if the velocity is positive or negative, it only gives you the change in velocity (increase or decrease). Finally, note that the velocity increases on all intervals where the acceleration is positive and the velocity decreases on all intervals where the acceleration is negative.

## GRAPHICAL RELATIONSHIP BETWEEN VELOCITY AND POSITION

Figure 3.12 shows the relationship between the area of a velocity interval and the change in position for the same time interval. In the first interval, the area of velocity is positive and has an area of 0.8 m. As a result, the position increases by 0.8 m over the same time interval. Knowledge of the area of velocity does not tell you if the position is positive or negative, it only gives you the change in position (increase or decrease). Finally, note that the position increases on all intervals where the velocity is positive and the position decreases on all intervals where the velocity is negative.

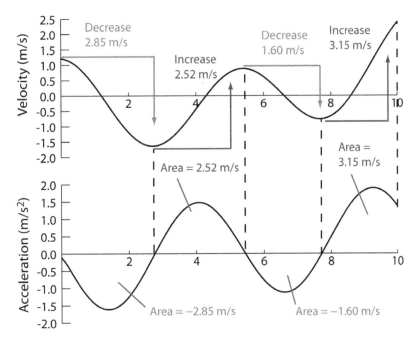

**Figure 3.11** The relationship between the area of an acceleration–time curve and the corresponding change in velocity. Notice that the velocity decreases on over all intervals where the acceleration is negative and the velocity increases over all areas where the acceleration is positive. The numerical area of each acceleration phase is directly equal to the change in velocity over the corresponding time interval.

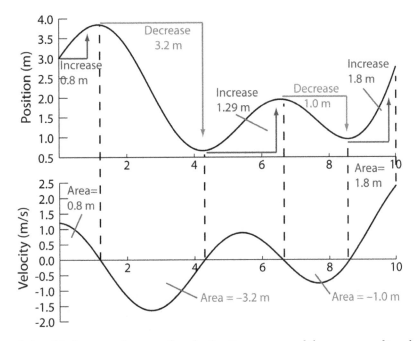

**Figure 3.12** The relationship between the area of a velocity–time curve and the corresponding change in position. Notice that the position decreases on over all intervals where the velocity is negative and the position increases over all areas where the velocity is positive. The numerical area of each velocity phase is directly equal to the change in position over the corresponding time interval.

## What does the Iinitial Value Do in Integration?

The integration of acceleration gives the change in velocity. The integration of velocity gives the change in position. The general form of the integration equation we have been using is:

$$P_f = \int (V \times dt) + P_i$$
$$P_f = Height \times Width + Initial\ Value$$
$$V_f = \int (A \times dt) + V_i$$
$$V_f = Height \times Width + Initial\ Value$$

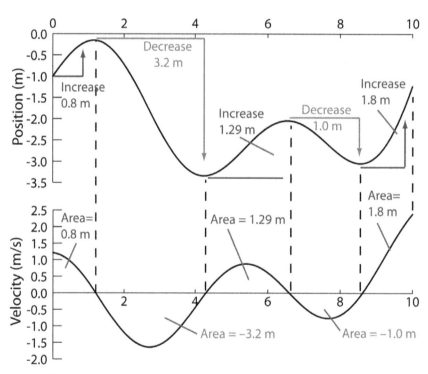

**Figure 3.13** The initial value for position is −1.0 m. Compare this position–time curve with the position–time curve shown in Figure 3.11. In Figure 3.11 the initial value for position is +3.0 m.

Hopefully by now you are wondering what the role of the initial value is when you integrate some function. It is actually very straightforward: the initial value for the function gives the y intercept. In other words, it tells you where to start drawing your graph at time = 0. Figure 3.13 contains the same curve for velocity, with one exception: the value for initial position is different. In Figure 3.13, the initial value for position is −1.0 m rather than +3.0 m as in Figure 3.11. Notice how the initial value simply moves the position curve up or down the y axis. This should clarify that integration gives a change in the graph above. A positive acceleration does not insure that the velocity will be positive, it only insures that the velocity will increase over the interval. Similarly, a negative acceleration does not indicate that the velocity is negative, it simply indicates that wherever velocity is (positive, zero, negative), it must go down or decrease.

## SUMMARY

This chapter defined position, velocity, and acceleration. The term "taking a derivative" was used to describe evaluating the slope of a position-time curve to determine the velocity or evaluating the slope of a velocity-time curve to determine the acceleration. Integration of an acceleration curve was used to compute the change in velocity over some time interval and integration of velocity was used to determine the change in position over some interval.

## REVIEW QUESTIONS

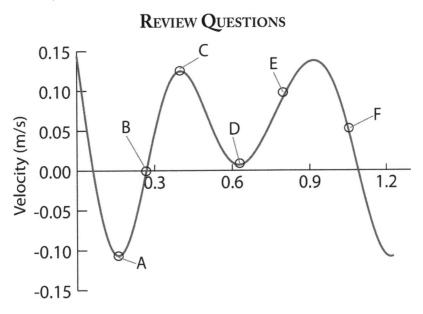

**Figure 13.14** Velocity-time graph used for review questions 1-15.

Use the velocity–time graph shown in Figure 3.14 to answer questions 1-15. Before answering these questions remember that you are looking at a velocity curve. For questions pertaining to position, you must evaluate the area and magnitude of velocity. For questions pertaining to acceleration, you must evaluate the slope of the velocity–time graph.

1. The position at point A is?
2. The acceleration at point A is?
3. The position at point B is?
4. The acceleration at point B is?
5. The position at point C is?
6. The acceleration at point C is?
7. The position at point D is?

8.  The acceleration at point D is?
9.  The position at point E is?
10. The acceleration at point E is?
11. The position at point F is?
12. The acceleration at point F is?
13. How will the negative velocity area affect position?
14. How will the positive velocity area affect position?
15. When you look at the entire velocity curve, are there more positive or negative areas, and how will this affect the change in position over the time interval from start t=0 to finish t= 1.2?

## Suggested Supplemental Resources

Cutnell, J. D. and K. W. Johnson (2008). Physics. Hoboken, John Wiley & Sons.

Enoka, R. M. (2008). Neuromechanics of Human Movement. Champaign, Human Kinetics.

Griffiths, I. W. (2006). Principles of Biomechanics & Motion Analysis. Philadelphia, Lippincott Williams & Wilkins.

Hall, S. J. (2007). Basic Biomechanics. Boston, McGraw Hill.

Hamill, J. and K. M. Knutzen (2009). Biomechanical Basis of Human Movement. Philadelphia, Lippincott Williams & Wilkins.

McGinnis, P. M. (2013). Biomechanics of Sport and Exercise. Champaign, Human Kinetics.

Thomas, G. B. (2004). Calculus, 11th ed. Boston, Addison Wesley.

# Chapter 4

# LINEAR KINEMATICS APPLICATIONS

## STUDENT LEARNING OUTCOMES

**After reading Chapter 4, you should be able to:**

- Graph position–velocity–acceleration curves in Excel.
- Use laboratory data to explain slope (derivatives) of data-time graphs.
- Use laboratory data to explain area (integration) of data-time graphs.
- Demonstrate how to compute derivatives (velocity & acceleration) in Excel.
- Demonstrate how to compute running integrals and areas of discrete regions in Excel.

## POSITION

Kinematic analysis of motion is performed by placing reflective markers over anatomical landmarks. Figure 4.1 shows an example of reflective marker placement. The motion of interest is filmed using two or more cameras. The x, y coordinates of each reflective marker from

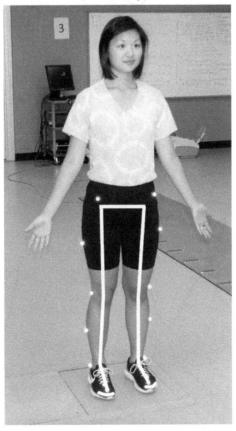

**Figure 4.1** Example of reflective marker placement for video analysis of motion for lower body gait analysis.

each camera view are then transformed into 3D (x, y, z) coordinates using computer software. Once 3D coordinates are obtained, the motion can be rotated and translated to view from any perspective. After obtaining the coordinates of the markers placed on the body surface, a series of mathematical transformations and rotations are performed in computer software to obtain anatomical joint centers in 3D space. The translations of coordinates use anatomical prediction equations to determine the x, y, z location of each body segment's joint center. The white lines drawn on the picture give a two–dimensional view of the lines connecting the anatomical joint

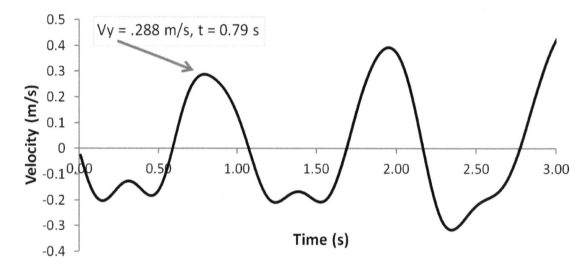

**Figure 4.2** Vertical velocity of the wrist during gait.

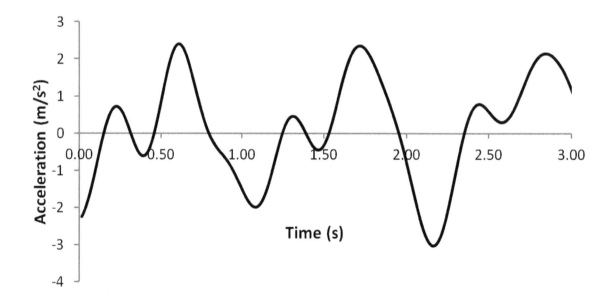

**Figure 4.3** Vertical acceleration of the wrist during gait.

centers. The computer output can include x, y, z coordinates of each marker, joint center, segment center of mass, and any other calculations that are performed.

## BASIC GRAPHING IN EXCEL

The examples in this section will use the Excel data file "**Sara Walk 5 XY Data.xls**", which is available on the course website. When you open the file, you will find several tabs at the bottom. Click on the Y Wrist tab and then create the following xy scatter graphs: Vy right wrist – time, and Ay right wrist – time. Your completed version of the xy scatter graph of the Vy right wrist velocity should look like the graph shown in Figure 4.2. When describing kinematic data, researchers usually describe phases of positive and negative values and the local maximum or minimum attained over the time interval. Here is an example. The vertical velocity of the right wrist was positive over the interval from 0.59–1.08 s and it reached a local peak value of .288 m/s at t = .79 s. The xy scatter plot for the vertical acceleration of the right wrist is shown in Figure 4.3.

### Two or More Plots on an XY Scatter Graph

Sometimes it is useful to plot two or more curves on the same graph. The graph shown in Figure 4.4 was created in Excel using an xy scatter plot. The wrist acceleration was plotted on the primary y axis and the wrist velocity was plotted on the secondary axis. Using a right and left y axis is useful when the functions to be plotted have a markedly different range of y values as in this case, where the acceleration ranges from −4 m/s² to +4 m/s² and the velocity ranges from −0.4 m/s to +0.4 m/s.

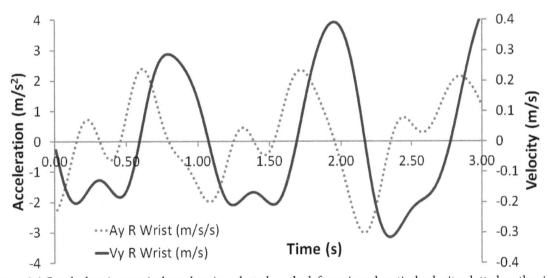

**Figure 4.4** Graph showing vertical acceleration plotted on the left y axis and vertical velocity plotted on the right y axis.

# COMPUTING DERIVATIVES IN EXCEL

## Velocity

Velocity is defined as the rate of change in position. The formula to compute velocity is

$$Velocity = \frac{\Delta Position}{\Delta Time}$$

or

$$V = \frac{P_f - P_i}{\Delta t}$$

Where $P_f$ is final position, and $P_i$ is initial position. You will see the above formula implemented in Excel in the Y Wrist tab. Look in cell C3 (the equation in cell C3 is shown below). All of the data in this file were filmed at 120 frames per second, so the time width is 1/120.

$$=(B3-B2)/(1/120)$$

## Acceleration

Acceleration is the rate of change in velocity, or the slope of the velocity-time graph. Acceleration is defined with the following formulas:

$$Acceleration = \frac{\Delta Velocity}{\Delta Time}$$

or

$$A = \frac{V_f - V_i}{\Delta t}$$

where $V_f$ is final velocity, and $V_i$ is initial velocity. Look in cell D4. The equation in D4 is:

$$=(C4-C3)/(1/120)$$

# INTERPRETATION OF DERIVATIVES

You should be able to look at a position-time curve and guess (estimate) what the velocity curve would look like. Similarly, you should be able to look at a velocity-time curve and guess (estimate) what the acceleration curve would look like. Examine the relationship between the slope of the velocity graph and the corresponding acceleration graph shown in Figure 4.5.

First, notice the horizontal tangent lines on the velocity graph. The horizontal tangent lines show the location in time for horizontal slopes in velocity. Notice that the acceleration for each of these horizontal slopes is 0 m/s². Next, the upward arrows indicate regions where the slope of the velocity is positive. The corresponding acceleration is always some positive number and the steeper the slope, the higher the positive acceleration. Finally, the downward

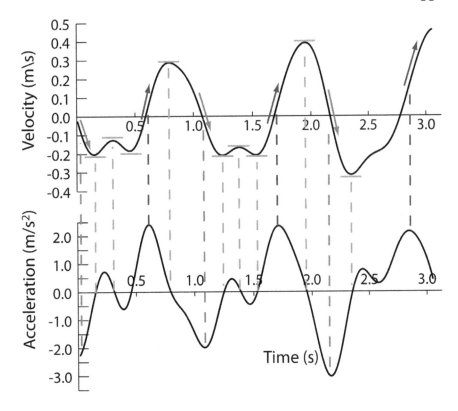

**Figure 4.5** Vertical velocity and acceleration of the wrist during gait.

arrrows indicate regions where the slope of the velocity is negative. The corresponding acceleration is always some negative number and regions a with steeper the slope have a greater magnitude of negative acceleration. The steepest spot on the entire curve occurs at a time of 2.17 seconds, where the acceleration is −3.009 m/s².

## INTERPRETATION OF AREAS TO ESTIMATE SLOPE OF VELOCITY OR POSITION

### The Effects of Negative Acceleration Areas on the Change in Velocity

Figure 4.6 depicts the relationship between negative acceleration and the corresponding decrease in velocity over the same time intervals. To interpret the effects of the area of an acceleration graph on the change in velocity, you must be able to think backwards. In other words, for any interval where the acceleration is negative (shown in gray in Figure 4.6), the velocity must have a negative slope, so the velocity must decrease. The value (magnitude) of the velocity may or may not be negative, but the slope of the velocity must be negative. Look at the first negative interval in the graph shown in Figure 4.6 from 0 to .15 seconds. The slope of the velocity is negative and the magnitude is also negative.

The second negative interval is similar to the first interval. The slope of the velocity is negative and the magnitude is negative. In the third negative interval (0.8 – 1.24 s) the slope of the velocity is negative, but the magnitude is not always negative. The magnitude is positive at the start of the interval and is negative at the end of the interval.

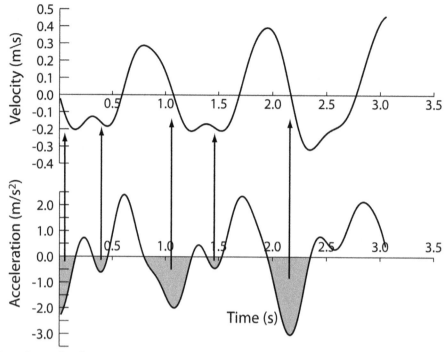

**Figure 4.6** The relationship between a negative acceleration (gray areas) and the resulting decrease in velocity over the corresponding time intervals.

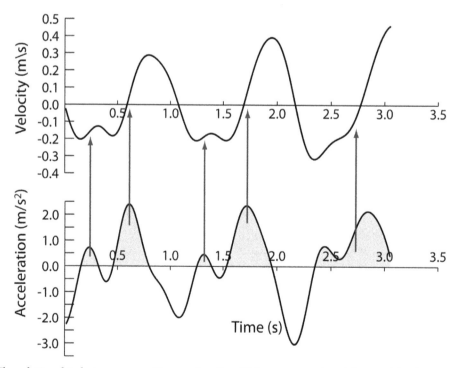

**Figure 4.7** The relationship between a positive acceleration (light gray areas) and the resulting increase in velocity over the corresponding time intervals.

In the fourth negative acceleration interval, both the slope of the velocity and the magnitude of the velocity are negative.

In the last negative acceleration interval from 1.96–2.35 s the slope of the velocity is negative. Like the third interval, the magnitude of the velocity is initially positive and then becomes negative. So now, here is the rule that holds: if the acceleration is negative, the slope of the velocity must be negative. The magnitude of the velocity can be either positive, negative, or zero over the interval.

### The Effects of Positive Acceleration Areas on the Change in Velocity

For any interval where the acceleration is positive (shown in light gray in Figure 4.7), the velocity must have a positive slope. The value (magnitude) of the velocity may be positive, negative, or zero, but the slope of the velocity must be positive. If the acceleration is positive, the slope of the velocity must be positive. The magnitude of the velocity can be either positive, negative, or zero over the interval.

## RUNNING INTEGRAL VS INTEGRAL WITH UPPER & LOWER LIMITS

The integration algorithm presented in chapter 3 is sometimes referred to as a running integral. The output gives a continuous velocity (or position) curve as a function of time. To obtain the change in velocity from an acceleration–time curve, the integration would be written as follows:

$$\Delta V = \int A \, dt$$
$$V_f - V_i = \int A \times dt$$
$$V_f = \int (A \times dt) + V_i$$

In the "**Sara Walk 5 XY Data.xls**" file, use the center of mass Ay data, located in the Running Integral tab. To implement the above equation, enter the following equation in cell C4 (Note, the initial velocity is already given in cell C3).

=(B3 * (1/120)) + C3

Copy this equation down the column C. This output in column C is a continuous velocity curve. Integration of the Ay for the center of mass (CM) would give the velocity curve shown in Figure 4.8.

## INTEGRATION OF INTERVALS TO GET AREA

Sometimes it is necessary to integrate intervals or regions of a curve. In the graph shown in Figure 4.9, each of the positive (shown in light gray) and negative (shown in gray) areas of the acceleration curve are integrated. The area of each acceleration interval represents the change in velocity for the given interval. For example, the area of the first positive acceleration interval (light gray) is +0.077 m/s. As a result, the velocity changes by +0.077 m/s. The

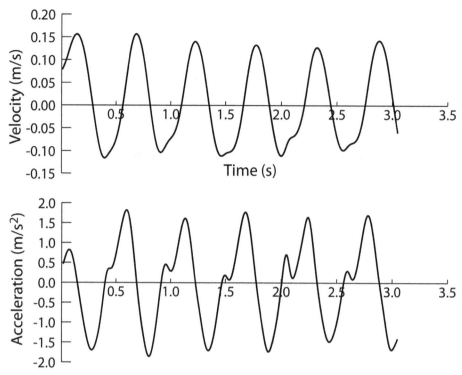

**Figure 4.8** Integrating the acceleration using a running integral gives a continuous velocity – time curve.

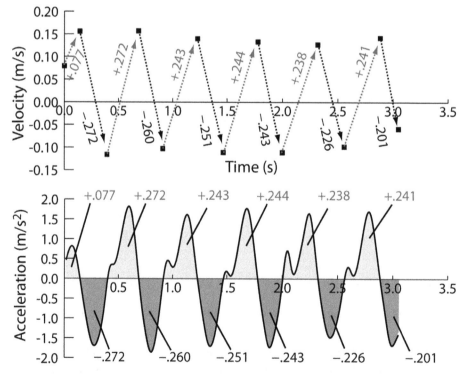

**Figure 4.9** Integration of positive and negative areas of an acceleration graph gives the corresponding change in velocity over the time interval integrated.

second interval of acceleration is negative and has an area of −0.272 m/s, and therefore the velocity will decrease by −0.272 m/s.

Now examine the Integration Pos and Neg Areas tab in the Excel file "**Sara Walk 5 XY Data.xls**". If you scroll down the column, you will see that the beginning and end of each positive and negative phase for Ay has been highlighted. The table at the top contains the start and end of each positive and negative phase.

The integration is handled in steps. The first step is to create a column to multiply Ay × dt (column C). In cell C4, enter the equation below, which computes the area of a rectangle ,Ay × dt (height × width). The value for acceleration is the height in units of m/s² and the value for time is in units of seconds. As a result, when you multiply acceleration by time, you are left with units of velocity m/s.

= B4 * (1/120)

Now to get the area for each interval, use the generic formula:

= sum (column-row: column-row)

To get the area of the first interval in column C from rows 4–19 enter the following equation in cell H3:

= SUM(C4:C19)

Now you should be able to follow the same logic to get the area of the remaining positive and negative intervals. The change in velocity is shown in column D. To get the change in velocity at the end of each interval, just add the velocity in the row immediately before the start of the interval to the area computed for the interval. The general formula is:

Initial velocity + area

Here is the exact equation that gives the velocity in cell D19:

=H3+D3

Complete this process to get the velocity at the end of each interval. In cell D49, the equation is:

=L3+D19

## COMPARISON OF RUNNING INTEGRAL TO INTEGRATING AREAS

Both the running integral method and the integrating areas method give the change in velocity. The difference between the two methods is the running integral gives a continuous velocity curve as a function of time, whereas the integrate areas method only gives the velocity at the end of each area. Look at the tab labeled "**Running vs Areas**" for a comparison of the two outputs. You should observe that the velocity matches at each point.

## COMPUTING VELOCITY AND POSITION FROM ACCELERATION

When given acceleration versus time, it is possible to compute velocity from acceleration using integration. Figure 4.10 shows an Excel file that contains acceleration–time data. Velocity-time data can be obtained by computing a running integral of the acceleration data. Position-time data can be obtained from the velocity-time data after computing velocity from acceleration.

### Computing Velocity from Acceleration

To compute velocity from acceleration' it is necessary to integrate acceleration to get the change in velocity using the formula below:

$$V_f = \int (A \times dt) + V_i$$

The general form of this equation is:

Final Velocity = (Acceleration × change in time) + Initial Velocity

To compute the velocity for cell C4 in the Excel worksheet shown above, enter the equation shown below in cell C4 and copy it down the C column.

=(B4*(1/120))+C3

### Computing Position from Velocity

To compute position from velocity, it is necessary to integrate velocity to get the change in position using the formula below:

$$P_f = \int (V \times dt) + P_i$$

The general form of this equation is:

Final Position = (Velocity × change in time) + Initial Position

To compute the position for cell D3 in the Excel worksheet shown above, enter the equation shown below in cell D2 and copy it down the D column:

=(C3*(1/120))+D2

## Summary

This chapter further elaborates on derivatives and integration of functions. The difference between integrating a region of a curve to get the area is contrasted with running integrals, which give a continuous curve as a function of time.

## Review Exercises

Download the file "**Lesson 4 Practice File.xls**", which is available on the course website. In "**Lesson 4 Practice File.xls**" complete the following:

1. Compute the velocity and acceleration from position, on the tab labeled "Get Deriv X CM".
2. Compute the velocity and position from acceleration, on the tab labeled "Integrate Ay".
3. Identify the start and end rows of all positive and negative phases for acceleration, on the tab labeled "Pos & Neg Areas of Ay", and then compute the area of each positive and negative phase.
4. Identify the location in time of the steepest upward phase of velocity on the tab labeled "Vel Steep Up & Down".
5. Identify the location in time of the steepest upward and downward phase of position on the tab labeled "Pos Steep Up & Down".

You can check your work with the file "**Lesson 4 Practice File ANSWERS.xls**".

## Suggested Supplemental Resources

Cutnell, J. D. and K. W. Johnson (2008). Physics. Hoboken, John Wiley & Sons.

Enoka, R. M. (2008). Neuromechanics of Human Movement. Champaign, Human Kinetics.

Griffiths, I. W. (2006). Principles of Biomechanics & Motion Analysis. Philadelphia, Lippincott Williams & Wilkins.

Hall, S. J. (2007). Basic Biomechanics. Boston, McGraw Hill.

Hamill, J. and K. M. Knutzen (2009). Biomechanical Basis of Human Movement. Philadelphia, Lippincott Williams & Wilkins.

McGinnis, P. M. (2013). Biomechanics of Sport and Exercise. Champaign, Human Kinetics.

Thomas, G. B. (2004). Calculus, 11th ed. Boston, Addison Wesley.

# Chapter 5

## PROJECTILES

### STUDENT LEARNING OUTCOMES

**After reading Chapter 5, you should be able to:**

- Describe the horizontal and vertical motion of a projectile.
- Demonstrate how to solve projectile motion problems.

### INTRODUCTION

When an object or performer is in the air, the horizontal acceleration (Ax) is 0 m/s² (if we ignore the effects of air resistance) and the vertical acceleration (Ay) is −9.8 m/s². Examples of projectiles include a ball, javelin, discus, long jumper, triple jumper, and diver. The horizontal velocity of the projectile is constant throughout the flight, since Ax = 0 m/s². The vertical velocity is affected by the force of gravity. For the upward portion of the flight, Vy will decrease, at the peak of the flight Vy = 0 m/s and on the downward portion of the flight, Vy will increase in the negative direction. At equal heights, the vertical velocity is equal in magnitude and opposite in direction. Figure 5.1 shows the parabolic trajectory of a projectile. The vertical arrows in the Figure 5.1 show the effect of gravity on vertical velocity (Vy). At takeoff, Vy is 12 m/s, at the peak Vy is 0 m/s, and at landing Vy is −12 m/s. The horizontal arrows show the horizontal velocity of the projectile, notice that the horizontal velocity is constant at 10 m/s

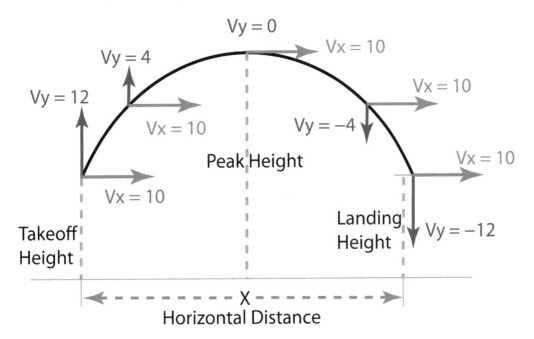

**Figure 5.1** Example flight of a projectile.

throughout the flight. Since we are ignoring the effects of air resistance, the horizontal velocity will be constant throughout the flight of the object or performer.

## VARIABLES USED TO DESCRIBE PROJECTILE MOTION

The variables used to solve projectile problems are displayed in Figure 5.2. Vertical velocities are shown with vertical arrows, heights are shown using dashed lines, and horizontal velocity distance is shown using horizontal arrows.

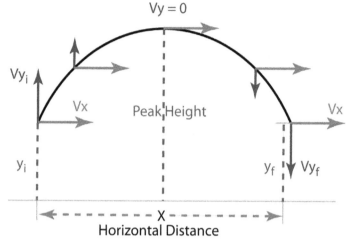

**Figure 5.2** Projectile motion variables.

### Definition of Variables

a = −9.8 m/s$^2$ (acceleration due to gravity)
t is time
$Vy_i$ is initial vertical velocity
$Vy_f$ is final vertical velocity
$y_i$ is initial height
$y_f$ is final height
x is horizontal distance or range
Vx is horizontal velocity

### Vertical Motion Equations

[1] $$Vy_f = Vy_i + at$$

[2] $$(Vy_f)^2 = (Vy_i)^2 + 2a(y_f - y_i)$$

[3] $$y_f = y_i + Vy_i(t) + \frac{1}{2}at^2$$

### Horizontal Motion Equation

[4] $$x = Vx(t)$$

## Using Vertical Motion Equations

Equations 1 thru 3 apply only to vertical motion. They are used to find vertical velocity, time and height. Equation 4 applies only to horizontal motion. It is used to find the time to cover some horizontal distance, the horizontal distance, or horizontal velocity.

Use equation [1] to solve for time (t), initial vertical velocity ($Vy_i$), or final vertical velocity ($Vy_f$) at some time:

[1]
$$Vy_f = Vy_i + at$$

Use equation [2] to solve for initial vertical velocity ($Vy_i$), or final vertical velocity ($Vy_f$) at some height, or the initial height ($y_i$), or final height ($y_f$). When using equation [2] to solve for a vertical velocity (Vy) that is going down, it will not give a negative number because both of the velocities are squared. So if you're solving for the vertical velocity, and you know that the object is going down, then you will have to change your answer to negative.

[2]
$$(Vy_f)^2 = (Vy_i)^2 + 2a(y_f - y_i)$$

Use equation [3] to solve for initial height ($y_i$) or final height ($y_f$) at a specific time (t).

[3]
$$y_f = y_i + Vy_i(t) + \frac{1}{2}at^2$$

## Horizontal Motion Equation

Use equation [4] to find the horizontal distance or range (x), or the time (t) to cover some horizontal distance.

[4]
$$x = Vx(t)$$

## Time in the Air and Vertical Velocity

The time in the air, or hang time for a projectile can be broken into two or more phases: time up and time down. The relationship between time in the air and the vertical motion of a projectile is shown in Figure 5.3. If the projectile takes off and lands at the same height,

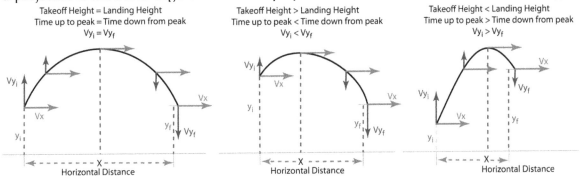

**Figure 5.3** Relationship between time in the air, take off and landing height, and vertical velocity.

shown on the left in Figure 5.3, then the time up to the peak is equal to the time down from the peak. In the middle graph of Figure 5.3, the object takes off at a higher height than the landing height. In this instance, the time up is less than the time down and the final vertical velocity ($Vy_f$) will be greater than the initial vertical velocity ($Vy_i$). Finally, as shown on the right of Figure 5.3, if the object takes off at a lower height than the landing height, then the time up is greater than the time down.

At the peak of the flight, $Vy = 0$. Many of the projectile problems will require that you remember that at the peak, the vertical velocity ($Vy$) is 0 m/s. Any time you read a problem that references the peak, remind yourself that at the peak, $Vy = 0$ m/s. Then you just have to decide if it is the $Vy_i$ or $Vy_f$ that is 0. For example, if the problem asks you to find the time to fall from the peak to the ground, then $Vy_i = 0$ m/s, because the object is beginning at the peak. Conversely, if the problem asks you to find the time to go from takeoff to the peak, then $Vy_f = 0$ m/s, because the object is beginning at takeoff and finishing at the peak.

Be sure to use symmetry of $Vy$ whenever it is helpful. In other words, at equal heights, the vertical velocity is equal and opposite. Therefore, if a ball has a vertical velocity of 8 m/s when it is 2 meters above the ground, it will have a vertical velocity of −8 m/s on the way down when it is the same height of 2 meters above the ground.

## The Terms Initial and Final are Relative

When solving problems on projectile motion, the terms initial and final are relative. Figure 5.4 gives several examples of initial and final conditions for projectile motion. As shown in Figure 5.4, initial and final can refer to different time points of the motion. While the terms initial and final can refer to different time points in the motion, initial will always occur in time before final. In Figure 5.4 drawing A, the initial height ($y_i$) and initial vertical velocity ($Vy_i$) represent takeoff and final height ($y_f$) and final vertical velocity ($Vy_f$) represent the landing conditions. In drawing B, the initial height ($y_i$) and initial vertical velocity ($Vy_i$) represent takeoff and final height $y_f$ and final vertical velocity ($Vy_f$) represent the peak where the final vertical velocity $Vy_f = 0$ m/s. In drawing C, the initial height ($y_i$) and initial vertical velocity ($Vy_i$) represent start of the flight with $Vy_i = 0$ m/s and final height ($y_f$) and final vertical velocity $Vy_f$ represent the landing conditions. In D, the initial conditions represent takeoff

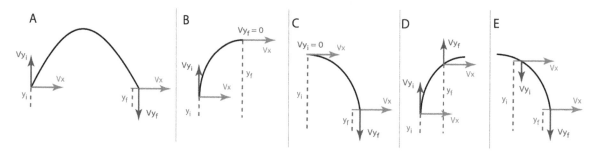

**Figure 5.4** The terms initial and final are relative when used in projectile motion problems. A) initial represents takeoff and final represents landing, B) initial represents takeoff and final represents the peak, C) initial represents the peak and final represents landing, D) initial and final represent the ascent, E) initial and final represent descent .

and the final conditions represent somewhere in the ascent of the flight. Finally in E, the initial conditions represent somewhere in the descent of the flight and final conditions represent the landing.

## EXAMPLE PROJECTILE PROBLEMS

### Equation 1: Finding the Time to Peak

A ball is kicked with an initial vertical velocity of 16 m/s. Find the time to the peak (see Figure 5.5). Note: at the peak, the vertical velocity is 0, so $Vy_f = 0$ m/s, and we always know a = $-9.8$ m/s².

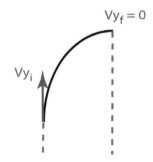

$$Vy_f = Vy_i + at$$
$$0 = 16 - 9.8(t)$$
$$9.8(t) = 16$$
$$t = \frac{16}{9.8}$$
$$t = 1.63s$$

Figure 5.5 Find time to peak.

### Equation 1: Finding the Velocity at a Specific Time

A ball is dropped with an initial vertical velocity $Vy_i$ of 0 m/s. Find the vertical velocity of the ball after 3 seconds (see Figure 5.6.)

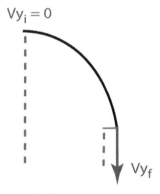

$$Vy_f = Vy_i + at$$
$$Vy_f = 0 - 9.8(3)$$
$$Vy_f = -29.4m/s$$

Figure 5.6 Find $Vy_f$ at t = 3 sec.

### Equation 2: Finding the Vertical Velocity without Time

A ball is thrown with an initial height ($y_i$) of 1.2 m and an initial vertical velocity ($Vy_i$) of 12 m/s. Find the final vertical velocity ($Vy_f$) of the ball when the height ($y_f$) is 3.6 m (see Figure 5.7).

$$(Vy_f)^2 = (Vy_i)^2 + 2a(y_f - y_i)$$

$$(Vy_f)^2 = (12)^2 + 2(-9.8)(3.6 - 1.2)$$

$$(Vy_f)^2 = 144 - 19.6(2.4)$$

$$(Vy_f)^2 = 144 - 47.07$$

$$\sqrt{(Vy_f)^2} = \sqrt{96.96}$$

$$Vy_f = 9.85 m/s$$

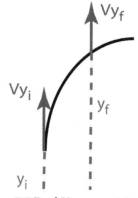

**Figure 5.7** Find $Vy_f$ at $y_f$ = 3.6 m.

### Equation 2: Finding the Final Vertical Velocity without Time

A ball falls from a peak height ($y_i$) of 24.69 m and lands with a final height ($y_f$) of 0 m. Determine the vertical velocity ($Vy_f$) of the ball at landing. Note: at the peak, either the initial or the final vertical velocity must be 0 m/s. In this case, the initial vertical velocity ($Vy_i$) = 0 m/s (see Figure 5.8).

$$(Vy_f)^2 = (Vy_i)^2 + 2a(y_f - y_i)$$

$$(Vy_f)^2 = (0)^2 + 2(-9.8)(0 - 24.69)$$

$$(Vy_f)^2 = 0 - 19.6(-24.69)$$

$$(Vy_f)^2 = 483.924$$

$$\sqrt{(Vy_f)^2} = \sqrt{483.924}$$

$$Vy_f = 22.00 m/s$$

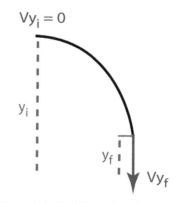

**Figure 5.8** Find $Vy_f$ at landing.

Since both of the velocities in equation 2 are squared, it will not give you a negative number for the answer. We know that the velocity should be negative in this problem, so we change the sign of the answer. The correct answer is:

$$Vy_f = -22.00 m/s$$

## Equation 3: Finding the Height at a Specific Time

A football is punted with an initial height ($y_i$) of 1.4 m and an initial vertical velocity ($Vy_i$) of 18 m/s. Find the height of the football ($y_f$) after 3 seconds (see Figure 5.9).

$$y_f = y_i + Vy_i(t) + \frac{1}{2}at^2$$

$$y_f = 1.4 + 18(3) + \frac{1}{2}(-9.8)(3)^2$$

$$y_f = 1.4 + 54 + (-4.9)(9)$$

$$y_f = 1.4 + 54 - 44.1$$

$$y_f = 11.3m$$

**Figure 5.9** Find $y_f$ at t = 3 sec.

## Equation 4: Finding the Horizontal Distance Traveled

A diver takes off with an initial vertical velocity ($Vy_i$) of 3.6 m/s, a horizontal velocity ($Vx$) of 1.2 m/s and an initial height $y_i$ of 2.2 m. Find the time for the diver to reach a peak and then fall back down to the same height as takeoff 2.2 m. Then determine the horizontal distance traveled during this time interval. As shown in Figure 5.10, you can use the fact that the diver takes off and lands at the same height to determine the final vertical velocity ($Vy_f$) is −3.6 m/s. Now find the time between these two velocity time points.

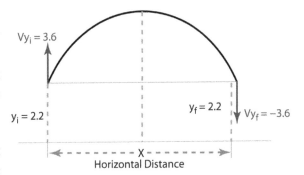

Figure 5.10 Determine the time between the two vertical velocity points and the horizontal distance traveled.

$$Vy_f = Vy_i + at$$

$$-3.6 = 3.6 - 9.8(t)$$

$$9.8t = 3.6 + 3.6$$

$$t = \frac{7.2}{9.8}$$

$$t = 0.73s$$

Now use equation [4] to find the horizontal velocity between these two time points.

$$x = Vx(t)$$

$$x = 1.2m/s(.73s)$$

$$x = 0.88m$$

# Review Concepts

1. Draw a picture of a projectile that takes off and lands at the same height. Show the horizontal and vertical velocity at the following time points: takeoff, midway on the ascent, at the peak, midway on the descent, at landing. Describe the relationship between time up, time down, and the total time in the air.

2. Draw a picture of a projectile that takes off at a height of 14 m and lands at a height of 0 m. Show the horizontal and vertical velocity at the following time points: takeoff, midway on the ascent, at the peak, midway on the descent, at landing. Describe the relationship between time up, time down, and the total time in the air.

3. A golf ball is hit at an initial height of 0 m and takes off with a vertical velocity of 18 m/s and a horizontal velocity of 27 m/s. What is the horizontal velocity of the ball at the instant that it strikes the ground?

# Review Problems

Correct answers are given in [square brackets]

1. A ball is thrown with a vertical velocity ($Vy_i$) of 13 m/s, horizontal velocity (Vx) of 15 m/s and an initial height ($y_i$) of 1.2 m.
   a. Find the height of the ball after 2 seconds. [$y_f$ = 7.6 m]
   b. Find the vertical velocity at 2 seconds [$Vy_f$ = –6.6 m/s]
   c. Find the horizontal distance covered in 2 seconds [x = 30 m]

2. A triple jumper reaches a peak height ($y_f$) of 2.1 m. If her takeoff height ($y_i$) was 1.1 m what was her takeoff vertical velocity? [$Vy_i$ = 4.43 m/s]

3. A football in the air has a height ($y_i$) of 27 m and a vertical velocity ($Vy_i$) of –2 m/s. What is the vertical velocity of the ball when it hits the ground (final height of 0 m)? [$Vy_f$ = –23.09 m]

4. A baseball leaves the bat with a vertical velocity ($Vy_i$) of 34 m/s and an initial height ($y_i$) of .8 m. How many seconds will it take for the ball to reach peak height ($y_f$)? [t = 3.47 s]

5. A soccer ball lands with a vertical velocity ($Vy_f$) of –13 m/s and a final height ($y_f$) of 0 m. What was the peak height of the ball? [$y_i$ = 8.61 m]

6. It's a windy day at Fenway Park in Boston. A Red Sox player hits a baseball with an initial height ($y_i$) of 0.9 m with a horizontal velocity (Vx) of 23 m/s and an initial vertical velocity ($Vy_i$) of 25 m/s. What horizontal distance did the ball travel if it lands with a final height ($y_f$) 0 m? [x = 118.22 m]

# Suggested Supplemental Resources

Cutnell, J. D. and K. W. Johnson (2008). Physics. Hoboken, John Wiley & Sons.

Enoka, R. M. (2008). Neuromechanics of Human Movement. Champaign, Human Kinetics.

Griffiths, I. W. (2006). Principles of Biomechanics & Motion Analysis. Philadelphia, Lippincott Williams & Wilkins.

Hall, S. J. (2007). Basic Biomechanics. Boston, McGraw Hill.

Hamill, J. and K. M. Knutzen (2009). Biomechanical Basis of Human Movement. Philadelphia, Lippincott Williams & Wilkins.

McGinnis, P. M. (2013). Biomechanics of Sport and Exercise. Champaign, Human Kinetics.

Thomas, G. B. (2004). Calculus, 11th ed. Boston, Addison Wesley.

# Chapter 6

## ANGULAR KINEMATICS

### STUDENT LEARNING OUTCOMES

**After reading Chapter 6, you should be able to:**

- Define angular kinematics, position, velocity, and acceleration.
- Define absolute and relative angles.
- Analyze the slope of an angular position curve to estimate angular velocity.
- Analyze the slope of an angular velocity curve to estimate angular acceleration.
- Compute the area of an angular acceleration curve to determine the change in velocity.
- Compute the area of an angular velocity curve to determine the change in position.
- Define the relationship between linear and angular velocity.
- Define tangential acceleration, centripetal acceleration, and centripetal force.
- Solve angular kinematics problems.

### ANGULAR MOTION

Angular motion is an integral component of human movement. When a muscle shortens, it causes a body segment to rotate about a joint center. Many sport motions, such as diving, involve rotation of the body about the center of mass.

Kinematics is defined as the description of motion. Angular kinematics is the description of angular motion. In angular kinematics, angular motion is described using angular position (angle), angular velocity, and angular acceleration.

### UNITS OF MEASUREMENT

Angular motion can be described using radians, degrees, or revolutions. The international standard unit for angular measurement is the radian. Many of the equations used for angular kinematics require that radians be used for angular measures. Figure 6.1 shows the definition of a radian. A radian is the ratio of the arc length (s) to the radius (r), shown in Figure 6.1. One radian is equal to 57.29 degrees, $\pi$ radians is equal to 180 degrees. As shown in the lower left of Figure 6.1, when the angle is 1 radian, the arc length (s) is equal to the radius (r). Since radians are a ratio of arc length to radius, they are a dimensionless unit and in some equations, the radians will disappear when multiplied by m, m/s or m/s$^2$.

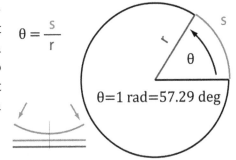

$$\theta = \frac{s}{r}$$

$\theta = 1 \text{ rad} = 57.29 \text{ deg}$

**Figure 6.1** A radian is the ratio of the arc length (s) to the radius (r).

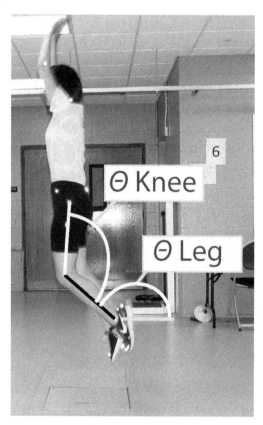

**Figure 6.2** A segment angle defines the angle of a body segment as measured counterclockwise from a right horizontal. A joint angle is the angle between any two body segments. The knee joint angle shown above is an example of a relative angle. The leg angle show above is an example of an absolute angle.

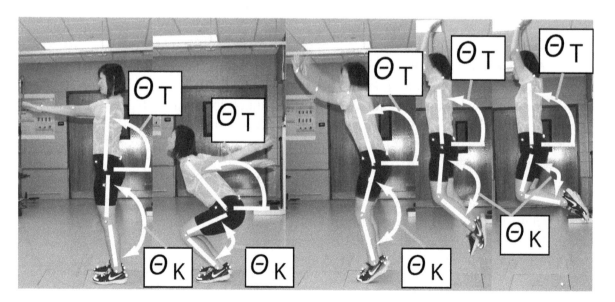

**Figure 6.3** Changes in trunk segment angles ($\theta_T$) and knee joint angles ($\theta_K$) during a vertical jump.

## ANGULAR POSITION

When analyzing human motion, absolute angles are used to describe body segment angles and relative angles are used to describe joint angles. The Greek symbol theta (θ) is traditionally used to describe an angle. An absolute angle, also referred to as a segment angle, is the measure of an angle to the right horizontal (see Figure 6.2). Absolute angles are usually measured in the counterclockwise direction from a right horizontal line. The right horizontal line can be placed at the proximal or distal end of a segment. A relative angle, also known as a joint angle, is the angle between two line segments. In Figure 6.2, the absolute angle for the leg and relative angle for the knee are shown. Figure 6.3 shows changes in trunk angles (absolute) and knee joint angles (relative) that occur during a standing vertical jump.

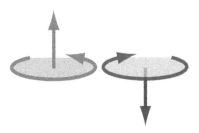

**Figure 6.4** The right hand rule is used to determine the direction of an angular motion vector. Curl the fingers of your right hand in the direction of rotation and your thumb will point in the direction of the vector.

## ANGULAR MOTION VECTORS

Vectors for angular position (θ), angular velocity (ω), and angular acceleration (α) are defined according to the right hand rule (see Figure 6.4). Counterclockwise rotation is positive (shown by upward directed arrow) and clockwise rotation is negative (shown by downward). To determine the direction of the vector, take you right hand and curl your fingers in the direction of the rotation and your thumb points in the direction of the vector. For the positive rotation, the angular vector points upward and for the negative rotation the angular vector points downward (see Figure 6.4).

## ANGULAR VELOCITY

Angular velocity is defined as the rate of change in angular position. The Greek symbol (ω) omega is used for angular velocity and the units for angular velocity are radians/sec (r/s). When the angular velocity is positive, the curved arrow is drawn in the counterclockwise direction (see Figure 6.5). The formula to compute angular velocity is:

$$Angular\ Velocity = \frac{\Delta\ Angular\ Position}{\Delta time}$$

or

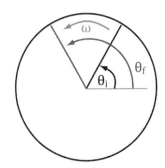

**Figure 6.5** Angular velocity is the rate of change in angular position.

$$\omega = \frac{\theta_f - \theta_i}{\Delta t}$$

Where $\theta_f$ is the final angle and $\theta_i$ is the initial angle.

## Sample Angular Velocity Problem

When kicking a soccer ball, the player's lower leg has an initial angle ($\theta_i$) of 0.2 r and just prior to contact with the ball, the players lower leg has a final angle ($\theta_f$) of 0.45 r. The time between these two angular positions was 0.14 s. Compute the angular velocity ($\omega$).

$$\omega = \frac{\theta_f - \theta_i}{\Delta t}$$

$$\omega = \frac{.45 - .2}{.14}$$

$$\omega = 1.79 r / s$$

Angular velocity can also be described as the slope of the angular position–time graph. In Figure 6.6, the slope of the position (angle) vs time is negative from t = 0 to t = .125 s, as a result, the angular velocity is negative over the interval. The slope of the position–time graph is horizontal at t = .125 s, and the corresponding angular velocity is 0 d/s at .125 s. Finally from t = .125 s to .3 s, the slope of the position–time graph is positive and the corresponding angular velocity is positive.

The angle–time and angular velocity–time graphs shown in Figure 6.6 were generated after filming a runner from a lateral view. The angle graph describes the subtalar joint pronation that occurs in running. Pronation is computed as the difference between the angle of the lower leg and the angle of the rear of the running shoe.

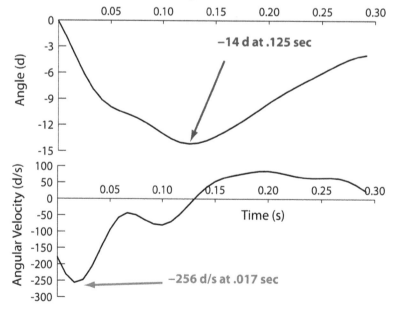

**Figure 6.6** Angular velocity is the rate of change in angular position.

## ANGULAR ACCELERATION

Angular acceleration is the rate of change in angular velocity or the slope of the velocity–time graph. The Greek symbol alpha (α) is used for angular acceleration. The units for angular acceleration are r/s². Angular acceleration is defined with the following formulas:

$$Angular\ Acceleration = \frac{\Delta\,Angular\ Velocity}{\Delta time}$$

or

$$\alpha = \frac{\omega_f - \omega_i}{\Delta t}$$

where $\omega_i$ is the initial angular velocity, $\omega_f$ is the final angular velocity and t is time.

### Sample Angular Acceleration Problem

A gymnast is swinging around the horizontal bar with an initial angular velocity ($\omega_i$) of 3.7 r/s and a radius of 1.5 m (See Figure 6.7). Compute the angular acceleration (α) necessary to increase his angular velocity ($\omega_f$) to 4.8 r/s in a time of 0.3 s.

$$\alpha = \frac{\omega_f - \omega_i}{\Delta t}$$

$$\alpha = \frac{4.8r/s - 3.7r/s}{.3s}$$

$$\alpha = 3.67r/s^2$$

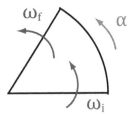

Figure 6.7 Angular acceleration is the rate of change in angular velocity.

## RELATIONSHIP BETWEEN LINEAR AND ANGULAR VELOCITY

The linear velocity ($v$) for an object rotating about some axis with a radius (r) is the product of the angular velocity ($\omega$) times the radius of rotation (r), (see Figure 6.8). The linear velocity vector ($v$) is tangent to the path of rotation and perpendicular to the radius. The equation for the relationship between linear and angular velocity is:

$$v = \omega r$$

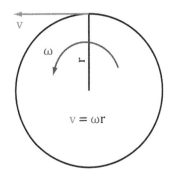

Figure 6.8 Relationship between linear and angular velocity.

where ω is the angular velocity in r/s, r is the radius in m, and v is the velocity in m/s. The angular velocity must be in units of r/s. Notice that when you multiply something in r/s by something in M, the result is m/s. When working with radians, you must remember they are a dimensionless unit since they are the ratio of the arc length to the radius. The relationship between linear and angular velocity is shown in Figure 6.8. For an object rotating at a constant angular velocity, increasing the radius will cause an increase the linear velocity.

In Figure 6.9, the angular velocity of the rigid rotating arm is constant at 2 r/s. Figure 6.9 illustrates the effect of increasing the radius from 2 m to 4 m on the resulting linear velocity. As shown in Figure 6.9, increasing the radius while keeping the angular velocity constant causes the linear velocity to increase from 4 m/s to 8 m/s.

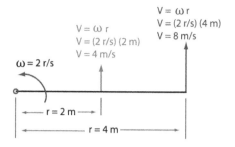

**Figure 6.9** Increasing the radius of rotation causes an increase in linear velocity.

### Sample Linear and Angular Velocity Problem

Calculate the linear velocity (*v*) obtained if a golfer swings a golf club with a radius (r) of 1.1 m using a angular velocity (ω) of 240 d/s. First, convert to radians. See Figure 6.10 for a diagram of the linear and angular velocity vectors.

$$radians = \frac{deg\,rees}{57.29}$$

$$\omega = \frac{240d/s}{57.29d/r}$$

$$\omega = 4.19r/s$$

Now solve for the linear velocity.

$$v = \omega r$$

$$v = (4.19r/s)(1.1m)$$

$$v = 4.16m/s$$

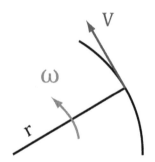

**Figure 6.10** Relationship between linear and angular velocity.

## LINEAR AND ANGULAR ACCELERATION

There are two linear accelerations experienced for an object rotating about some axis: tangential acceleration ($a_T$) and centripetal acceleration ($a_c$), see Figure 6.11.

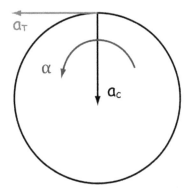

**Figure 6.11** Tangential acceleration ($a_T$) is tangent to the path, and centripetal acceleration ($a_c$) is directed inward toward the center of the circle. The units for tangential and centripetal acceleration are m/s².

## TANGENTIAL ACCELERATION

The tangential acceleration ($a_T$) for an object rotating about an axis is always tangent to the path and perpendicular to the radius. The units for tangential acceleration are m/s². If the angular acceleration is positive, then the tangential acceleration will be positive. If the angular acceleration is negative, then the tangential acceleration will be negative. If the angular acceleration is zero, then the tangential acceleration will be zero. There are two equations for computing tangential acceleration; both are shown in Figure 6.12. Tangential acceleration ($a_T$) is the product of angular acceleration and the radius of rotation (see the equation on the left in Figure 6.12). The equation on the right in Figure 6.12 shows the relationship between the change in linear speed for an object rotation about some axis and the tangential acceleration.

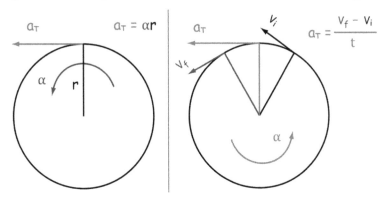

**Figure 6.12** Formulas for computing tangential acceleration.

### Sample Tangential Acceleration Problem 1

A golfer is swinging a 13 kg club with an initial linear velocity ($V_i$) of 5.7 m/s and a radius of 1.9 m about the shoulder axis (see Figure 6.13 for a diagram). After 0.6 seconds, the club has a final linear velocity ($V_f$) of 10.3 r/s. Find the tangential acceleration ($a_T$).

$$a_T = \frac{V_f - V_i}{t}$$

$$a_T = \frac{10.3 - 5.7}{0.6}$$

$$a_T = 7.67 m / s^2$$

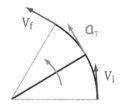

**Figure 6.13** Tangential acceleration.

### Sample Tangential Acceleration Problem 2

When throwing a 9 kg discus, a track athlete accelerates the discus with an angular acceleration (α) of −4.5 r/s² for a time (t) of 0.4 s with a radius (r) of 1.3 m (see Figure 6.14). Compute the tangential acceleration ($a_T$).

$$a_T = \alpha r$$

$$a_T = (-4.5r / s^2)(1.3m)$$

$$a_T = -5.85 m / s^2$$

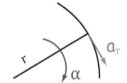

**Figure 6.14** Tangential acceleration.

## CENTRIPETAL ACCELERATION

An object rotating about some axis must always be accelerated inward toward the center to keep the object rotating about the axis. This acceleration necessary to keep an object rotating about some axis is called the centripetal acceleration ($a_c$). There are two equations for centripetal acceleration. Both are shown below in Figure 6.15. The units for centripetal acceleration are m/s².

The equation on the left shows the relationship between the angular velocity (ω), the radius of rotation (r), and the resulting centripetal acceleration ($a_c$). Since the angular velocity (ω) is squared, increasing the angular velocity has a much bigger effect on the centripetal acceleration than does increasing the radius.

The second equation for computing centripetal acceleration ($a_c$) is shown on the right. It relates linear velocity (v) and the radius (r). Like the equation on the left, the velocity com-

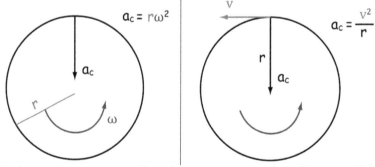

**Figure 6.15** Formulas for computing centripetal acceleration. Centripetal acceleration is directed inward toward the center of the circle.

ponent is squared so it has a larger effect on centripetal acceleration ($a_c$) than does the radius. It is important to note that the centripetal acceleration can never be zero for an object rotating about some axis. If the centripetal acceleration is zero, the object will leave the circle. Softball and baseball players often tie a ball to a string and spin it overhead to use for hitting practice. If the string breaks, the ball will leave the circle tangent to the path of the circle.

## CENTRIPETAL FORCE

The force necessary to keep an object rotating in a circle or about some axis is directly related to the centripetal acceleration. This force required to keep an object rotating about some axis is called centripetal force ($F_c$). The units for centripetal force are N. Its magnitude is simply mass times the centripetal acceleration ($a_c$). Like centripetal acceleration, centripetal force is directed inward toward the center of the circle or axis of rotation. The formulas for computing centripetal force are shown in Figure 6.16. Notice that both equations are simply mass times the centripetal acceleration ($a_c$).

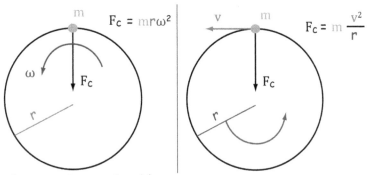

Figure 6.16 Formulas for computing centripetal force.

### Sample Centripetal Acceleration and Force Problem 1

A bowler rotates his arm with a radius of 1.2 m and a linear velocity (V) of 4.9 m/s. Compute the centripetal acceleration ($a_c$) and centripetal force ($F_c$) necessary to keep the 8 kg bowling ball in the circle (see Figures 6.17 & 6.18).

$$a_C = \frac{V^2}{r}$$

$$a_C = \frac{4.9^2}{1.2}$$

$$a_C = 20 m/s^2$$

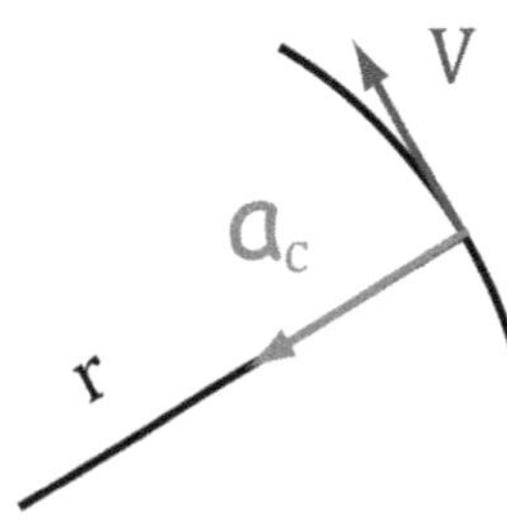

Figure 6.17 Centripetal acceleration.

$$F_C = m\frac{V^2}{r}$$

$$F_C = 8kg\frac{(4.9m/s)^2}{1.2m}$$

$$F_C = 160N$$

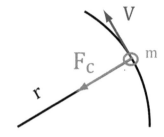

**Figure 6.18** Centripetal force.

## Sample Centripetal Acceleration and Force Problem 2

Compute the centripetal acceleration ($a_c$) and centripetal force ($F_c$) for a 70 kg runner going around a track turn with a radius (r) of 10 m and an angular velocity ($\omega$) of 2.3 r/s (see Figures 6.19 & 6.20).

$$a_C = r\omega^2$$

$$a_C = (10m)(2.3r/s)^2$$

$$a_C = 52.9m/s^2$$

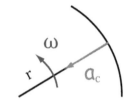

**Figure 6.19** Centripetal acceleration.

$$F_C = mr\omega^2$$

$$F_C = (70kg)(10m)(2.3r/s)^2$$

$$F_C = 3703N$$

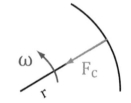

**Figure 6.20** Centripetal force.

## SUMMARY

Angular kinematics describes the position, velocity, and acceleration of objects rotating about some axis. The relationship between linear and angular velocity indicates that the linear velocity of a rotating object is a function of the radius and the angular velocity. The linear acceleration components, tangential and centripetal acceleration, are related to the rotational speed and the radius of rotation. Finally, centripetal force describes the force necessary for an object to rotate about some axis.

## ANGULAR KINEMATICS REVIEW QUESTIONS

1. A softball player throws the ball with an initial angular velocity ($\omega_i$) of 2.2 r/s, and after 0.6 seconds, the player's final angular velocity ($\omega_f$) is 3.3 r/s. Compute the angular acceleration ($\alpha$). Draw a picture of the angular velocities and the angular acceleration.
2. Calculate the linear velocity ($v$) obtained if a golfer swings a golf club with a radius of 1.2 m using a angular velocity ($\omega$) of 330 d/s.

3. Calculate the linear velocity ($v$) obtained if a golfer swings a golf club with a radius of 0.7 m using a angular velocity ($\omega$) of 330 d/s.

4. A javelin thrower's forearm has an initial angular velocity ($\omega_i$) of 6.3 r/s and a radius of 0.7 m. Compute the angular acceleration ($\alpha$) necessary to increase his angular velocity ($\omega_f$) to 6.8 r/s in a time of 0.3 s.

5. A discus thrower rotates his arm with an initial linear velocity ($V_i$) of 7.7 m/s and a radius of 1.2 m about the shoulder axis. After 0.6 seconds, the club has a final linear velocity ($V_f$) of 9.3 r/s. Find the tangential acceleration ($a_T$). Draw a picture showing the velocity components and the tangential angular velocity.

6. When throwing a 12 kg hammer, a track athlete accelerates ($\alpha$) the hammer −5.5 r/s² for a time of 0.4 sec with a radius (r) of 1.3 m. Compute the tangential acceleration ($a_T$). Draw a picture of the angular and tangential acceleration vectors.

7. A bowler rotates his arm with a radius of 1.3 m and a linear velocity (V) of 6.9 m/s. Compute the centripetal acceleration ($a_c$) and centripetal force ($F_c$) necessary to keep the 10 kg bowling ball in the circle. Draw a picture of the linear velocity, centripetal acceleration, and the centripetal force.

8. Compute the centripetal acceleration ($a_c$) and centripetal force ($F_c$) for a 80 kg runner going around a track turn with a radius (r) of 12 m and an angular velocity ($\omega$) of 1.9 r/s. Draw a picture of the linear velocity, centripetal acceleration and the centripetal force.

9. A runner contacts the ground with an initial lower leg angle ($\theta_i$) of 0.4 r, and after a time of 0.3 seconds, the final lower leg angle is ($\theta_f$) is −.25 r. Compute the angular velocity ($\omega$).

## SUGGESTED SUPPLEMENTAL RESOURCES

Cutnell, J. D. and K. W. Johnson (2008). Physics. Hoboken, John Wiley & Sons.

Enoka, R. M. (2008). Neuromechanics of Human Movement. Champaign, Human Kinetics.

Griffiths, I. W. (2006). Principles of Biomechanics & Motion Analysis. Philadelphia, Lippincott Williams & Wilkins.

Hall, S. J. (2007). Basic Biomechanics. Boston, McGraw Hill.

Hamill, J. and K. M. Knutzen (2009). Biomechanical Basis of Human Movement. Philadelphia, Lippincott Williams & Wilkins.

McGinnis, P. M. (2013). Biomechanics of Sport and Exercise. Champaign, Human Kinetics.

Thomas, G. B. (2004). Calculus, 11th ed. Boston, Addison Wesley.

# Chapter 7

## LINEAR KINETICS: REACTION FORCES

### STUDENT LEARNING OUTCOMES

**After reading Chapter 7, you should be able to:**

- Define kinetics, mass, inertia, and free-body diagrams.
- State Newton's Laws of Motion.
- Identify the difference between applied forces and reaction forces.
- Demonstrate how to draw free-body diagrams.
- Solve linear kinetics problems on reaction forces and accelerations.

### INTRODUCTION

**Newton's Laws of Motion**

In the seventeen century, Isaac Newton developed his three laws of motion by expanding upon the earlier work of his predecessors Copernicus, Galileo, and Kepler. Newton's laws of motion: (I. Law of Inertia, II. Law of Acceleration, and III. Law of Reaction), were published in his book **Philosphiae Naturalis Principia Mathematica**. Newton is often referred to as the father of calculus, mechanics, and planetary motion. **Kinetics**, which is a branch of mechanics, can be defined as the study of forces and the motion they cause.

There are three general approaches which can be used to solve kinetic problems (1) direct application of the Law of Acceleration, (2) application of impulse-momentum, and (3) application of work-energy. All three of these methods are derived from Newton's Law of Acceleration. In this chapter we will focus on kinetic problems using the direct application of the Law of Acceleration, commonly referred to as the force-mass-acceleration method. In chapters 8 and 9, we will use impulse-momentum to solve kinetic problems, and in chapter 10 we will use work - energy.

A **force** is defined as a push or a pull. Forces can be divided into contact forces and non-contact forces. Contact forces arise anytime two objects physically come into contact. Examples of contact forces include pushing against the ground, pushing a shot put, pulling or kicking in the water. Non-contact forces arise when one object exerts a force on another object without physically contacting it. Gravity is an example of a non-contact force. Even though the words force and mass are frequently used interchangeably, this is not correct. **Mass** is a scalar, it only has magnitude. The units for mass are kg. **Force** is a vector, it has both magnitude and direction. The units for force are Newtons (N). A **Newton** is defined as the amount of force necessary to accelerate a 1 kg mass at a rate of 1 m/s$^2$. **Inertia** is the natural tendency of an object to remain at rest or in motion at a constant speed in a straight line. Mass is a quantitative measure of the amount of inertia an object has. An object with greater mass will be more difficult to stop if it is moving or more difficult to move if it is at rest than an object with less mass.

**Law of Inertia**

I) **Law of Inertia** – A body at rest stays at rest, a body in motion stays in motion along a straight line unless acted upon by a net force. This law is also called the **law of motion**, as it refers to constant velocity and zero velocity.

The term net force is very important. It is not uncommon for a body to be acted upon by several forces. The net force is the vector sum of all of the forces acting on the body. In Figure 7.1, the horizontal and vertical forces cancel out, so the net horizontal and vertical forces are 0 N.

According to the **law of inertia** moving at a constant velocity is equivalent to being at rest (zero velocity). A net force would be required to change an object's state of motion, whether it is at rest or moving at a constant velocity. Since mass is the measure of the amount of inertia an object possesses, a greater net force would be required to change the motion of an object with a large mass, when compared to an object with less mass.

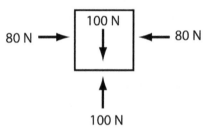

**Figure 7.1** Net forces acting on the box are zero.

II) **Law of Acceleration** – The acceleration experienced by a body is directly proportional to the net force and inversely proportional to its mass and it occurs in the direction of the force. The Law of Acceleration is explained by the following equation.

$$\Sigma F = ma$$

Where **F** is force in N, m is mass in kg, and **a** is acceleration in m/s$^2$. Figure 7.2A shows the effect of increasing the force on acceleration. Increasing the force from 50 to 100 N causes acceleration to increase from 5 m/s$^2$ to 10 m/s$^2$.

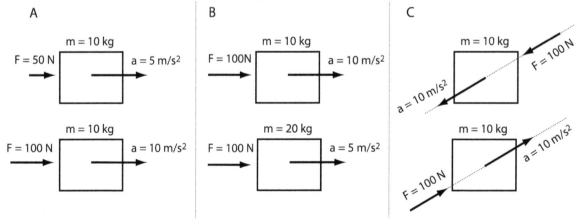

**Figure 7.2** (A) The effect of increasing the force upon the resulting acceleration. (B) The effect of increasing the mass on the resulting acceleration. (C) The acceleration occurs in the direction of the net force.

Figure 7.2B shows the effect of increasing the mass on the resulting acceleration when the net forces are constant. Increasing the mass from 10 kg to 20 kg decreases the acceleration from 10 m/s² to 5 m/s².

The last part of the **law of acceleration** states that the acceleration occurs in the direction of the force. In Figure 7.2C, the force vector depicts the direction of the net force. According to the Law of Acceleration the acceleration is always in the direction of the net force. Thus in Figure 7.4 the acceleration is in the same direction as the net force.

**III) Law of Reaction** – For every action, there is an opposite and equal reaction.

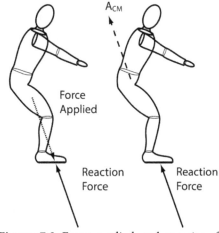

The **law of reaction** will be used to solve for forces acting on a system or performer. It is important to distinguish between applied forces and reaction forces. Figure 7.3 shows a performer landing. The performer applies a downward forward force against the ground, which is the applied force (dotted arrow in Figure 7.3). The applied force accelerates the ground in the direction of the force. The reaction force is equal in magnitude and opposite in direction to the applied force. The *reaction force accelerates the performer's center of mass in the direction of the reaction force* (see Figure 7.3).

**Figure 7.3** Force applied and reaction force during landing. The CM always accelerates in the direction of the reaction force.

## FREE-BODY DIAGRAMS

A free-body diagram (FBD) is a picture that represents an object or performer and all of the external forces acting on the object or performer. Free-body diagrams are used to determine either the force or the resulting acceleration when using Newton's **law of acceleration** to solve problems. Figure 7.4 shows someone lifting a box. To draw the FBD, begin by drawing in the force of gravity using downward arrow located at the center of mass of the object or performer to represent the force of gravity. Label this downward force vector "mg" or "wt". After drawing the force of gravity, examine the object or performer for any points of contact with the ground or some external object. For the lifter shown in Figure 7.4, the feet are in contact with the ground and the hands are in contact with the box. Draw in a positive x and y force for all points of contact with the ground or external objects. It is customary to label the forces under the feet as "Rx" for horizontal reaction force and "Ry" for vertical reaction force. The x and y forces acting on the hands due to contact with the box were labeled "Bx" for the x component of the box force and "By" for the y component of the box force. Always assume the directions of the contact forces are positive, so draw the y component upward and the x component to the right. When we solve for the actual forces using Newton's **law of acceleration** we should be able to determine if the forces are positive or negative.

In the FBD shown in Figure 7.4, the forces acting on the lifter were analyzed separately from the forces acting on the box. The box could have been included with the lifter simply by including the weight of the box with the weight of the lifter. The drawing in the middle in Figure 7.4 shows the forces acting on the lifter and the drawing on the right shows the forces acting on the box.

The drawing of the performer lifting the box in Figure 7.4 is a side-view drawing of the motion. While it is safe to assume that both feet are in contact with the ground and both hands are in contact with the box, you will notice that we only included a single point of contact for the feet and hands. You might be wondering why we didn't separate the forces acting on the feet into right and left foot and the forces acting on the hands into right and left hands. In this text, we will only include one point of contact for hands and feet for side-view drawings of the body. Thus, unless you are specifically told to find the forces acting on the right and left foot

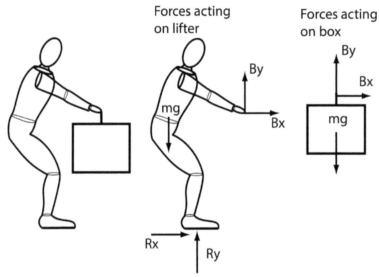

**Figure 7.4** A free–body diagram is a picture of all of the external forces acting on a system.

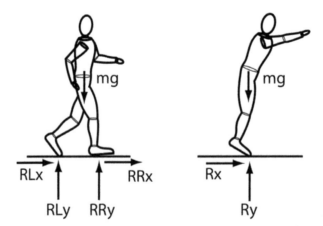

**Figure 7.5** Comparison of free-body diagrams including right and left foot with a side-view free-body diagram where the feet are modeled as a single point of contact.

just draw in one point of force contact for the feet and label it "Rx" and "Ry" for horizontal and vertical reaction forces as shown in Figure 7.4. When including both feet in a free body diagram you can label the x and y forces with numbers or letters. Figure 7.5 illustrates an example of a free body diagram that includes forces acting on the right and left foot and a side-view drawing where the feet are modeled using a single point of contact for the reaction forces.

## X-Y COMPONENTS FOR THE LAW OF ACCELERATION

The equation shown below for the **law of acceleration** is a vector equation. The force (F) and the acceleration (a) are printed in bold to indicate that they are vectors.

$$\Sigma Fx = ma_x \qquad\qquad \Sigma Fy = ma_y \qquad\qquad \Sigma Fz = ma_z$$

The vector equation above can be written using x, y, z components of the vectors as shown below:

$$\Sigma F = ma$$

In this course, we will generally focus on only the x and y versions. In words, the Fx equation above says: "add up all of the x (horizontal) forces and set them equal to mass times the x (horizontal) acceleration". Similarly, the Fy equation above says: "add up all of the y (vertical) forces and set them equal to mass times the y (vertical) acceleration".

## SOLVING FOR REACTION FORCES WHEN GIVEN ACCELERATION

In the next two problems, you will be given the accelerations and asked to find the reaction forces that caused the observed accelerations.

**Sample Problem 7-1**

The runner in Figure 7.6 has a mass of 60 kg. The runner experiences a horizontal acceleration ($a_x$) of −4.8 m/s² and a vertical acceleration ($a_y$) of 16 m/s². Complete the FBD and solve for the horizontal (Rx) and vertical (Ry) reaction forces that caused these accelerations. See Figure 7.6 for the solution. Note, g = −9.8 m/s² acceleration due to gravity.

$$\Sigma Fx = ma_x$$
$$Rx = ma_x$$
$$Rx = (60kg)(-4.8m/s^2)$$
$$Rx = -288N$$

$$\Sigma Fy = ma_y$$
$$Ry + mg = ma_y$$
$$Ry + (60kg)(-9.8) = (60kg)(16m/s^2)$$
$$Ry = 1548N$$

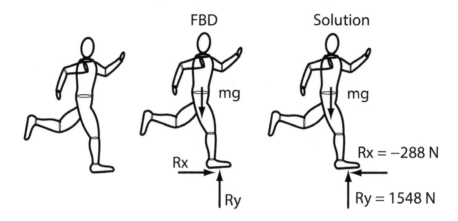

**Figure 7.6** Solving for reaction forces when given acceleration.

## Sample Problem 7-2

The diver in Figure 7.7 has a mass of 70 kg. The diver experiences a horizontal acceleration $a_x$ of 5.1 m/s² and a vertical acceleration $a_y$ of 15 m/s². Complete the FBD and solve for the horizontal (Rx) and vertical (Ry) reaction forces that caused these accelerations.

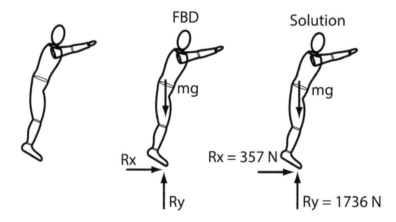

**Figure 7.7** Solving for reaction forces when given acceleration.

$$\Sigma Fx = ma_x$$
$$Rx = ma_x$$
$$Rx = (70kg)(5.1m/s^2)$$
$$Rx = 357N$$

$$\Sigma Fy = ma_y$$
$$Ry + mg = ma_y$$
$$Ry + (70kg)(-9.8) = (70kg)(15m/s^2)$$
$$Ry = 1736N$$

## Solving for Acceleration When Given Reaction Forces

In the next two problems, you will be given the reaction forces and asked to find the accelerations that the reaction forces caused.

### Sample Problem 7-3

When landing from a rebound, an 80 kg basketball player experiences a horizontal reaction force (Rx) of −300 N and a vertical reaction force (Ry) of 1700 N. Find the horizontal $(a_x)$ and vertical $(a_y)$ accelerations these reaction forces caused (see Figure 7.8). Notice in the solution that the direction of the accelerations is the same direction as the net forces. In addition, these reaction forces accelerate the basketball player's center of mass.

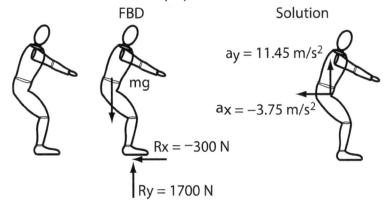

**Figure 7.8** Solving for acceleration when given reaction forces.

$$\Sigma Fx = ma_x$$
$$Rx = ma_x$$
$$-300N = (80kg)(a_x)$$
$$a_x = -3.75m/s^2$$

$$\Sigma Fy = ma_y$$
$$Ry + mg = ma_y$$
$$1700 + (80kg)(-9.8) = (80kg)(a_y)$$
$$a_y = 11.45\ m/s^2$$

### Sample Problem 7-4

The runner shown in Figure 7.9 has a mass of 75 kg. The runner experiences a horizontal reaction force (Rx) of 420 N and a vertical reaction force (Ry) of 1980 N. Complete the free-body diagram and determine the horizontal $(a_x)$ and vertical $(a_y)$ accelerations of the runner's center of mass caused by these reaction forces.

$$\Sigma Fx = ma_x$$
$$Rx = ma_x$$
$$420N = (75kg)(a_x)$$
$$a_x = 5.60\ m/s^2$$

$$\Sigma Fy = ma_y$$
$$Ry + mg = ma_y$$
$$1980 + (75kg)(-9.8) = (75kg)(a_y)$$
$$a_y = 16.60\ m/s^2$$

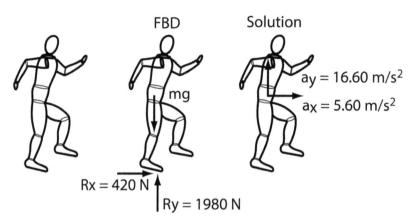

**Figure 7.9** Solving for acceleration when given reaction forces.

## SUMMARY

This chapter illustrated how Newton's Laws of Motion can be used to determine the forces that cause human movement. Using free-body diagrams, it is possible to determine the forces that cause an observed acceleration, or using forces, the accelerations they cause can be computed.

## REVIEW QUESTIONS

1. What is the difference between an applied force and a reaction force? Which force accelerates your center of mass, the applied force or the reaction force?
2. State Newton's three laws.
3. Draw a free-body diagram of a sprinter positioned in the blocks (both feet and both hands are in contact with the ground.
4. Draw a free-body diagram of a high jumper in the air.
5. Define inertia.
6. If the net reaction forces Rx and Ry acting on a runner are in the positive x direction and positive y direction, what will be the directions for $a_x$ and $a_y$ of the runner's center of mass?
7. An 85 kg runner experiences a horizontal acceleration ax of −5.9 m/s² and a vertical acceleration ($a_y$) of 18.6 m/s². Complete the FBD and solve for the horizontal (Rx) and vertical (Ry) reaction forces that caused these accelerations.
8. A 70 kg triple jumper experiences a horizontal acceleration ($a_x$) of 7.4 m/s² and a vertical acceleration ay of 14.3 m/s². Complete the FBD and solve for the horizontal (Rx) and vertical (Ry) reaction forces that caused these accelerations.
9. A 50 kg gymnast experiences a horizontal reaction force (Rx) of −520 N and a vertical reaction force (Ry) of 1210 N. Complete the free-body diagram and determine the horizontal ($a_x$) and vertical ($a_y$) accelerations of the gymnast's center of mass caused by these reaction forces.

## SUGGESTED SUPPLEMENTAL RESOURCES

Cutnell, J. D. and K. W. Johnson (2008). Physics. Hoboken, John Wiley & Sons.

Enoka, R. M. (2008). Neuromechanics of Human Movement. Champaign, Human Kinetics.

Griffiths, I. W. (2006). Principles of Biomechanics & Motion Analysis. Philadelphia, Lippincott Williams & Wilkins.

Hall, S. J. (2007). Basic Biomechanics. Boston, McGraw Hill.

Hamill, J. and K. M. Knutzen (2009). Biomechanical Basis of Human Movement. Philadelphia, Lippincott Williams & Wilkins.

Hibbeler, R. C. (2010). Engineering Mechanics: Dynamics. Upper Saddle River, Prentice Hall.

Hibbeler, R. C. (2010). Engineering Mechanics: Statics. Upper Saddle River, Prentice Hall.

McGinnis, P. M. (2013). Biomechanics of Sport and Exercise. Champaign, Human Kinetics.

Winter, D. A. (2009). Biomechanics and Motor Control of Human Movement. Hoboken, John Wiley & Sons, Inc.

# Chapter 8

## LINEAR KINETICS: IMPULSE–MOMENTUM

### STUDENT LEARNING OUTCOMES

**After reading Chapter 8, you should be able to:**

- Define impulse and momentum.
- Define the **principle of conservation of linear momentum**.
- Define elastic and inelastic collisions.
- Define the linear impulse-momentum relationship.
- Solve impulse-momentum problems.

### LINEAR MOMENTUM

The linear momentum ($p$) of an object is the product of the object's mass (m) and velocity ($v$) as follows:

$$p = mv$$

The vector for linear momentum points in the same direction as the velocity. The units for linear momentum are kg·m/s. An example linear momentum problem is shown in Figure 8.1. To determine the linear momentum for a 5 kg ball with a velocity of 3 m/S, multiply the mass times the velocity of the ball.

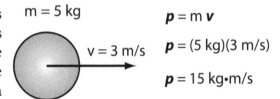

**Figure 8.1** Example linear momentum problem.

### PRINCIPLE OF CONSERVATION OF LINEAR MOMENTUM

The total linear momentum of a system of objects is constant if the net external force acting on the system is zero. The total linear momentum is defined by the following equation:

$$m_A u_A + m_B u_B = m_A v_A + m_B v_B$$

where m is mass, u is the velocity of the objects before the collision, and v is the velocity of the objects after the collision. Collisions are often classified according to whether the kinetic energy changes during the collisions. The two classifications used are **elastic** and **inelastic**. In an **elastic collision**, the total kinetic energy is the same before and after the collision. A good example of an elastic collision would be the collision between two billiard balls. The equation above can be rearranged to determine the velocity of each object after the collision:

$$v_A = \frac{m_B u_B}{m_A} \qquad\qquad v_B = \frac{m_A u_A}{m_B}$$

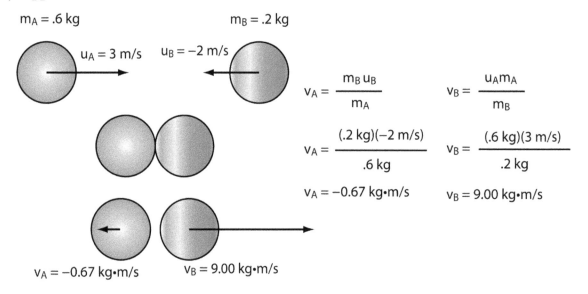

Figure 8.2 Example of conservation of linear momentum in an elastic collision.

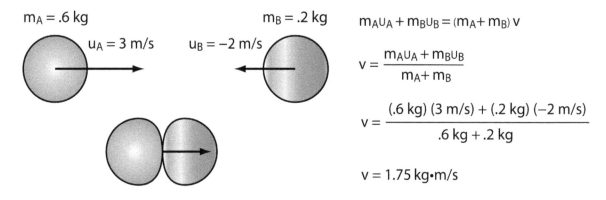

Figure 8.3 Example of conservation of linear momentum in an inelastic collision.

Figure 8.2 depicts an example of applying the **principle of conservation of linear momentum** for an elastic collision between two billiard balls.

### Sample Elastic Collision Problem

As shown in Figure 8.2, a 0.6 kg ($m_A$) billiard ball with an initial velocity ($u_A$) of 3 m/s collides with a 0.2 kg ($m_B$) billiard ball with an initial velocity ($u_B$) of −2 m/s in a perfectly elastic collision. Determine the velocity of each billiard ball after the collision.

In a perfectly inelastic collision, the linear momentum of the system is still conserved, but the two objects stick together and move together with the same velocity. The linear momentum equation can be rearranged to determine the velocity of combined objects after the collision as follows:

$$m_A u_A + m_B u_B = (m_A + m_B)v$$

Figure 8.3 depicts an example of applying the **principle of conservation of linear momentum** for an inelastic collision between two billiard balls.

## Sample Inelastic Collision Problem

As shown in Figure 8.3, 0.6 kg ($m_A$) piece of soft clay with an initial velocity ($u_A$) of 3 m/s collides with another piece of soft clay with a mass ($m_B$) of 0.2 kg and an initial velocity ($u_B$) of −2 m/s in a perfectly inelastic collision or plastic collision. Determine the velocity (v) of the pieces of clay after the collision (see Figure 8.3).

## IMPULSE-MOMENTUM

Forces that occur over very short intervals are called impulsive forces. Examples of impulsive forces include the collision of a golf club with a golf ball, collision of a foot with a soccer ball, and the collision of a bat with a ball. In each case, the force acting on the object is not constant, it varies with time. Figure 8.4 shows a force that varies with time. The product of the average force and the time of contact is called the impulse. Impulse is defined as the product of the average force and the time interval ($\Delta t$) during which the force acts. Impulse (J) is a vector quantity, as it has the same direction as the average force and the units for impulse are N·s. Linear impulse can be computed using average force and the time of force application as follows:

$$J = \bar{F}\Delta t$$

where J is impulse in N·s, $\bar{F}$ is the average force and $\Delta t$ is the time which the force acts. Multiplying the average force by time is to computes the area underneath the force-time curve which has units of area (N·s). Since impulse is the area under the force-time curve, the following equation also gives the impulse:

$$J = \int_{t_0}^{t_1} F \, dt$$

where the impulse (J) is computed by determining the area underneath the force-time curve by integrating force with respect to time.

### Sample Linear Impulse Problem Using Average Force

Use the average force impulse equation to compute the impulse for the force-time curve shown in Figure 8.4.

$$J = \bar{F}\Delta t$$
$$J = (95.6N)(.217s)$$
$$J = 20.7N \cdot s$$

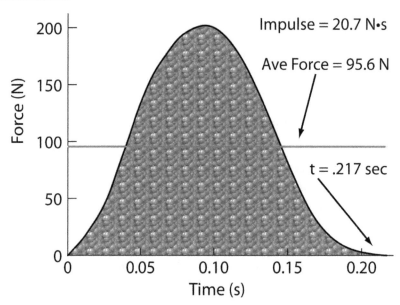

**Figure 8.4** Impulse can be computed using average force x time or integrating the force with respect to time.

Newton's second law can be used to explain the relationship between impulse and momentum as follows. First recall the formula for computing acceleration.

$$A = \frac{V_f - V_i}{\Delta t}$$

And now here is Newton's second law:

$$F = ma$$

Substituting the acceleration equation for *a* in F = m a gives the following.

$$F = m\left(\frac{V_f - V_i}{\Delta t}\right)$$

Now we can rearrange the equation to:

$$F(\Delta t) = mV_f - mV_i$$

This equation is the impulse-momentum relationship. Impulse is on the left side of the equation and the right side of the equation is the change in linear momentum. This equation indicates that the average force times time or the area under the force-time curve is equal to the change in linear momentum.

Some collisions involve relatively high forces applied over a relatively short time interval, such as the impact of a golf club with a golf ball or a shoe with a soccer ball. Figure 8.5 illustrates how the impulse-momentum equation can be used to compute the impulse applied to a soccer ball during a soccer kick. Measuring impulse directly for an impact between a foot and a soccer ball can be difficult. It would require a force device such as a load cell or a strain gauge to be mounted on the shoe and then the force would have to be sampled at a rate of 1,000-10,000 samples per second to accurately measure the impulse. An alternative solution is to use the right side of the impulse momentum relationship. Since the impulse is equal to the change in momentum, to indirectly measure the impulse one would need to determine the mass of the object and the velocity of the object before and after the collision.

## Sample Impulse-Momentum Problem

A soccer player imparts the impulsive force shown below on a soccer ball with a mass of 0.43 kg. Prior to the collision of the foot with the soccer ball, the ball has an initial velocity ($V_i$) of 0.0 m/s. After the collision, the ball has a final velocity ($V_f$) of 23.02 m/s. The soccer player imparts an average force of 90.8 N for a time of 0.109 s. Determine the impulse using the following methods:

1. Average force times time.
2. Change in linear momentum.

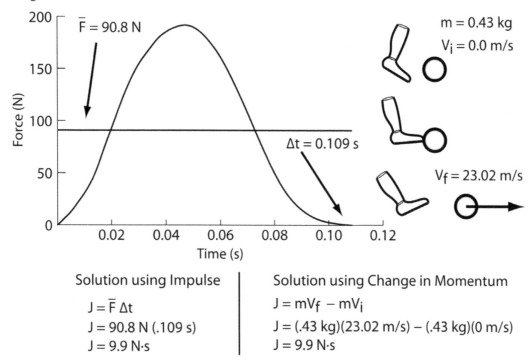

Solution using Impulse

$$J = \bar{F}\,\Delta t$$
$$J = 90.8\ \text{N}\ (.109\ \text{s})$$
$$J = 9.9\ \text{N·s}$$

Solution using Change in Momentum

$$J = mV_f - mV_i$$
$$J = (.43\ \text{kg})(23.02\ \text{m/s}) - (.43\ \text{kg})(0\ \text{m/s})$$
$$J = 9.9\ \text{N·s}$$

**Figure 8.5** Impulse can be computed using average force × time (left), or the change in momentum (right).

As shown in Figure 8.5, impulse can be computed either by measuring the force and the time of force application, or by measuring the change in velocity before and after the force is applied and computing the change in momentum.

$$\text{Average Force times Time} = \text{Change in Linear Momentum}$$
$$\bar{F}(\Delta t) = mV_f - mV_i$$

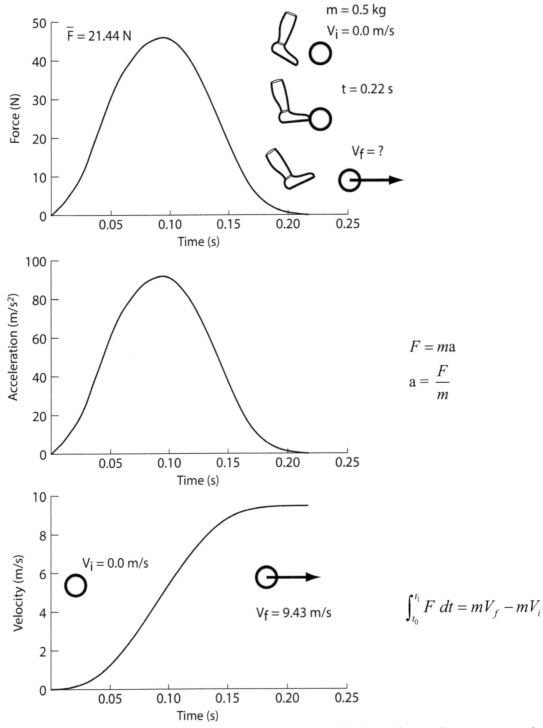

**Figure 8.6** Relationship between force, acceleration, and velocity. The shape of an acceleration curve is always the same as the force curve. Dividing the force by mass gives acceleration in units of m/s². Given a force-time curve, the corresponding velocity-time curve can be computed using the impulse-momentum relationship.

The Area under the Force-Time Curve = Change in Linear Momentum

$$\int_{t_0}^{t_1} F \, dt = mV_f - mV_i$$

## FORCE-ACCELERATION-VELOCITY RELATIONSHIP

When you are given a force-time curve, you should be aware that the corresponding acceleration-time curve has the exact same shape; the only difference between the two curves is that they have different units, the force is in units of N, and the acceleration is in m/s². Using Newton's **law of acceleration** and dividing the force by mass gives the corresponding acceleration for any force-time curve. Figure 8.6 shows a force-time curve and the corresponding acceleration-time curve and velocity-time curve. Notice that the acceleration has the exact same shape as the force curve. Now that you know that, you can look at a force-time curve and immediately know what the acceleration time curve would look like. It should give you the ability to predict the velocity-time curve that would result from any force-time curve. Recall from chapter 3 that integration of acceleration gives the change in velocity. If the acceleration is positive, the velocity must increase, if the acceleration is negative, the velocity must decrease. The velocity will be steep upward when the acceleration is at a local max and steep downward when the acceleration is at a local min. The velocity-time curve in Figure 8.6 can be computed directly from force using the impulse-momentum relationship or it can be computed by integrating the acceleration curve.

Using the average force of 21.44 N, a time of 0.22 s, and a mass of 0.5 kg for the soccer ball shown in Figure 8.6, the final velocity of the soccer ball can be computed as follows:

$$\bar{F}(\Delta t) = mV_f - mV_i$$
$$21.44N(.22s) = 0.5kg(V_f) - 0.5kg(0m/s)$$
$$V_f = 9.43m/s$$

## EFFECT OF IMPULSE ON VELOCITY

If two force-time curves have the same impulse but different force-time values, the final velocity of an object will be the same (this assumes that the object has the same initial velocity and mass). Figure 8.7 illustrates this concept. The dotted line force curve has a much higher force and shorter time than the solid line force curve. In each case, the ball has an initial velocity of 0 m/s and a mass of 0.5 kg. Since both curves have the same impulse, after each is applied to the ball, it will have the same final velocity of 9.6 m/s. The corresponding velocity-time curves for each force curve are shown in Figure 8.7. Examination of the velocity-time curves shows that the ball has the same final velocity after each force curve is applied to the ball. The final velocity is the same but the manner in which the ball attains this final velocity is markedly different between the two force curves. Since the dotted line force curve has a much higher magnitude peak force it causes a much greater acceleration of the ball. After applying the dotted line force curve, the ball reaches the final velocity of 9.6 m/s in 0.5 seconds. The magnitude of the peak force for the solid line force curve is much lower, and as a result, it causes a lower acceleration of the ball. The ball now takes 1.0 s to attain the same final velocity of 9.6 m/s.

To summarize this effect, if two force curves have the same impulse, they will cause the same final velocity on an object, assuming that the initial velocity and mass are the same in

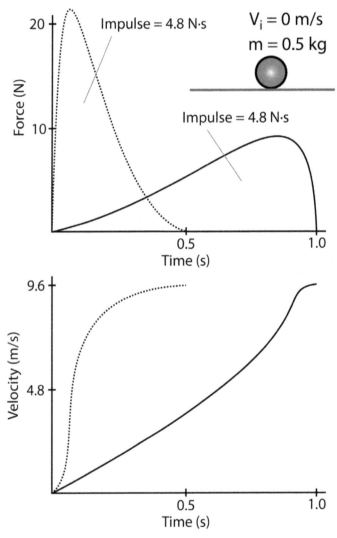

**Figure 8.7** The effect of force-time curves with the same impulse. The two force-time curves shown in the force graph at the top are each applied to a ball with an initial velocity of 0 m/s and a mass of 0.5 kg. The impulse for each force curve is 4.8 N·s. The resulting velocity-time curve for each force is shown in the bottom graph. In each case, the final velocity of the ball is 9.6 m/s. When the dotted line force curve is applied to the ball, it takes 0.5 s for the ball to attain the final velocity of 9.6 m/s. The solid line force curve has a lower magnitude, which then causes a lower acceleration of the ball. For the solid line force curve, it takes 1.0 s for the ball to attain the final velocity of 9.6 m/s.

each case. The final velocity is the same, but the slope of the velocity-time curve and the time to reach the final velocity will be different.

## Summary

This chapter is the first of two chapters on impulse-momentum. As outlined in this chapter, impulse is either average force times time or the area underneath the force-time curve. According to the impulse-momentum relationship, the impulse is equal to the change in momentum. Finally, momentum is the product of mass times velocity.

## Review Questions

1. Define linear momentum. What are the units for linear momentum? How is the direction of the vector determined?
2. Define impulse. What are the units for impulse? How is the direction of the impulse vector determined? Give two methods for computing impulse.
3. State the **principle of conservation of linear momentum**.
4. What is the difference between an elastic collision and an inelastic collision?
5. Compute the linear momentum for a 12 kg ball that has a velocity of 4.7 m/s.
6. A 0.4 kg ($m_A$) billiard ball with an initial velocity ($u_A$) of 2.3 m/s collides with a 0.2 kg ($m_B$) billiard ball with an initial velocity ($u_B$) of −2.9 m/s in a perfectly elastic collision. Determine the velocity of each billiard ball after the collision.
7. A 1.6 kg ($m_A$) piece of soft clay with an initial velocity ($u_A$) of 4.2 m/s collides with another piece of soft clay with a mass ($m_B$) of 0.9 kg and an initial velocity ($u_B$) of −2.3 m/s in a perfectly inelastic collision. Determine the velocity (v) of the pieces of clay after the collision.
8. State the impulse-momentum equation.
9. An average force of 85 N is applied for 0.15 seconds to a 1.2 kg ball. Prior to applying the force the ball has an initial velocity ($V_i$) of 0 m/s. Determine the impulse applied, the change in momentum, and the final velocity ($V_f$) of the ball.

## Suggested Supplemental Resources

Cutnell, J. D. and K. W. Johnson (2008). Physics. Hoboken, John Wiley & Sons.

Enoka, R. M. (2008). Neuromechanics of Human Movement. Champaign, Human Kinetics.

Griffiths, I. W. (2006). Principles of Biomechanics & Motion Analysis. Philadelphia, Lippincott Williams & Wilkins.

Hall, S. J. (2007). Basic Biomechanics. Boston, McGraw Hill.

Hamill, J. and K. M. Knutzen (2009). Biomechanical Basis of Human Movement. Philadelphia, Lippincott Williams & Wilkins.

Hibbeler, R. C. (2010). Engineering Mechanics: Dynamics. Upper Saddle River, Prentice Hall.

Hibbeler, R. C. (2010). Engineering Mechanics: Statics. Upper Saddle River, Prentice Hall.

McGinnis, P. M. (2013). Biomechanics of Sport and Exercise. Champaign, Human Kinetics.

Winter, D. A. (2009). Biomechanics and Motor Control of Human Movement. Hoboken, John Wiley & Sons, Inc.

# Chapter 9

## LINEAR IMPULSE–MOMENTUM APPLICATIONS

### STUDENT LEARNING OUTCOMES

**After reading Chapter 9, you should be able to:**

- Define horizontal impulse–momentum.
- Define vertical impulse–momentum.
- Define braking and propulsion impulses.
- Solve impulse–momentum problems.

### IMPULSE–MOMENTUM RELATIONSHIP

Impulse can be computed either by measuring the force and the time of force application, or by measuring the change in velocity before and after the force is applied and computing the change in momentum.

$$\text{Average Force} \times \text{Time} = \text{Change in Linear Momentum}$$
$$\bar{F}(\Delta t) = mV_f - mV_i$$

The Area under the Force–Time Curve = Change in Linear Momentum

$$\int_{t_0}^{t_1} F \, dt = mV_f - mV_i$$

The two equations above are general equations for computing impulse. They do not differentiate between horizontal and vertical impulse. In this chapter we will differentiate between horizontal and vertical impulse and then apply the above equations to running, walking, and the jumping force–time data.

### HORIZONTAL FORCE AND HORIZONTAL IMPULSE IN RUNNING

**Braking Impulse**

Figure 9.1 shows horizontal force and horizontal velocity for a runner during ground contact. In addition, the free-body diagram is shown for three phases of the motion: heel-strike, mid-stance and toe-off. Begin by examining the horizontal force from t = 0 to t = .112 s, notice that the horizontal reaction force is negative. In the FBD for heel-strike, the horizontal force (Rx) is shown as negative. Since this force (Rx) is opposing the runner's motion, it will cause the runner to decelerate. Also notice that the Rx force causes the runner's center of mass to have a negative horizontal acceleration ($a_x$). In biomechanics, it is customary to refer to the impulse caused by this force which opposes motion as a braking impulse. The result of this braking impulse is to decrease the horizontal velocity of the runner's center of mass from Vx

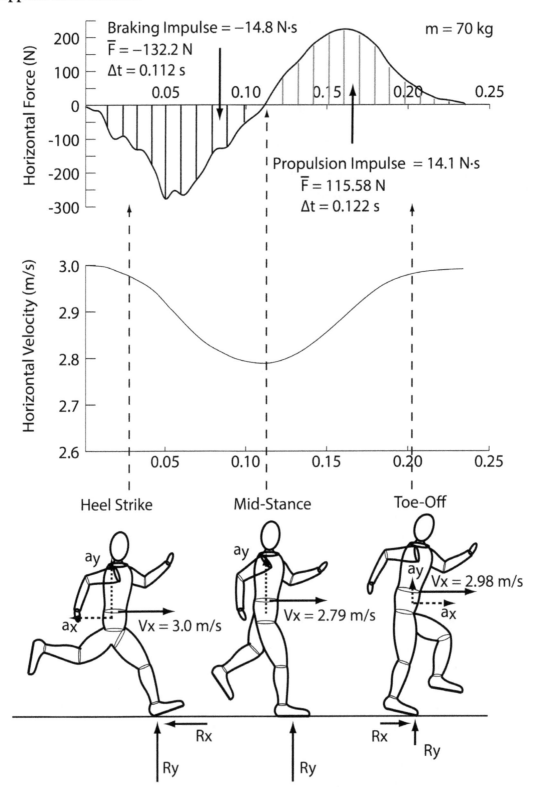

**Figure 9.1** Relationship between horizontal force, horizontal impulse, and the change in horizontal velocity.

= 3.0 m/s at heel-strike to Vx = 2.79 m/s at mid-stance.   A propulsion impulse is defined as an impulse where the direction of the impulse is the same as the direction of motion.  In the horizontal force graph below, the force Rx is positive from t = .113 to t = .235 s. The propulsion impulse from this positive force causes the runner's horizontal velocity to increase from mid-stance to toe-off.

## Example 1

Use the average force and area under the curve impulse-momentum formulas to compute the $Vx_f$ for the runner shown in Figure 9.1.  Average force is shown below.

$$\bar{F}(\Delta t) = mV_f - mV_i$$

$$\bar{Rx}(\Delta t) = mVx_f - mVx_i$$

$$-132.2N(.112s) = 70kg(Vx_f) - 70kg(3.0\,m/s)$$

$$Vx_f = 2.79\,m/s$$

Integration of Rx with respect to time to obtain the area underneath the force-time curve is shown below.

$$\int_{t_0}^{t_1} F\,dt = mV_f - mV_i$$

$$\int_0^{.112} Rx\,dt = 70kg(Vx_f) - 70kg(3.0\,m/s)$$

$$-14.7N \cdot s = 70kg(Vx_f) - 70kg(3.0\,m/s)$$

$$Vx_f = 2.79\,m/s$$

After experiencing the braking impulse, the runner's center of mass has slowed down to 2.79 m/s.  Next, we will now compute the impulse for the propulsion phase.

## Propulsion Impulse

The propulsion phase begins at mid-stance at t = .113 s.  At the start of propulsion, the horizontal velocity is 2.79 m/s, so the final velocity from the previous step (braking phase) becomes the initial velocity for propulsion. The net force is positive, so the acceleration must be positive. Computation of the propulsion impulse using the average force formula is shown below.

$$\bar{Rx}(\Delta t) = mVx_f - mVx_i$$

$$115.58N(.122s) = 70kg(Vx_f) - 70kg(2.79\,m/s)$$

$$Vx_f = 2.99\,m/s$$

Computation of the propulsion impulse using integration of the horizontal force is shown below.

$$\int_{t_0}^{t_1} F\,dt = mV_f - mV_i$$

$$\int_{.113}^{.235} Rx\,dt = 70kg(Vx_f) - 70kg(2.79\,m/s)$$

$$14.1N \cdot s = 70kg(Vx_f) - 70kg(2.79\,m/s)$$

$$Vx_f = 2.99\,m/s$$

## VERTICAL FORCE AND VERTICAL IMPULSE IN RUNNING

When computing vertical impulse, it is necessary to account for the vertical impulse caused by the force of gravity. Figure 9.2 shows the free-body diagram for a runner. Substituting the kinematic formula for vertical acceleration into Newton's second law of motion gives the vertical impulse-momentum equation. Similar to the general equation for impulse-momentum, vertical impulse-momentum can be computed using either the average vertical force or integrating to get the area under the vertical force-time curve.

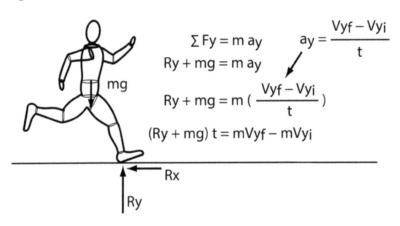

Figure 9.2 Vertical impulse momentum in running.

Figure 9.3 shows vertical force and vertical velocity for a runner during ground contact. In addition, the free-body diagram is shown for three phases of the motion: heel-strike, mid-stance, and toe-off. The vertical force (Ry) is positive throughout ground contact. At heel-strike the vertical velocity of the center of mass ($Vy_i$) is negative ($Vy_i = -.6$ m/s). Since the force of Ry is opposing the runner's motion, it will cause the vertical velocity of the runner's center of mass to change from negative ($Vy_i = -.6$ m/s) at heel-strike to positive ($Vy_f = .48$ m/s) at toe-off. The first peak in the vertical force of just over 1500 N that occurs at 0.026 s is referred to as an impact peak. A good running shoe should attenuate this impact peak before passing the load onto the lower leg. The ankle, knee, and hip joint stiffness of the runner at impact can also increase or decrease the impact peak. If the runner lands with increased joint stiffness, it will increase the impact peak. If the runner lands with decreased joint stiffness, it will decrease the impact peak. Therefore the impact peak is affected by the runner's shoe and joint stiffness.

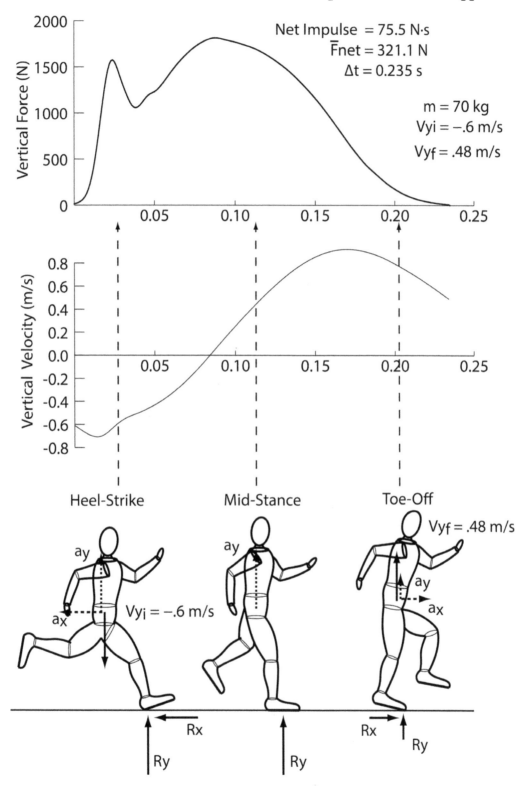

**Figure 9.3** Vertical force, vertical impulse and horizontal velocity during running.

### Example 2

Use the average force and area under the curve impulse-momentum formulas to compute the Vyf for the runner below.

### Average Force x Time is Equal to the Change in Linear Momentum

The example shown below is using the average net vertical force to compute the final vertical velocity ($Vy_f$).

$$(\bar{R}y + mg)(\Delta t) = mVy_f - mVy_i$$

$$321.1N(.235s) = 70kg(Vy_f) - 70kg(-.6\,m/s)$$

$$Vy_f = 0.48\,m/s$$

### The Area Under the Force-Time Curve is Equal to Change in Linear Momentum

This example is using the integration of the vertical force to compute the vertical impulse and the final vertical velocity ($Vy_f$).

$$\int_{t_0}^{t_1} F\,dt = mV_f - mV_i$$

$$\int_0^{.235} (Ry + mg)\,dt = 70kg(Vy_f) - 70kg(-.6\,m/s)$$

$$75.5N \cdot s = 70kg(Vy_f) - 70kg(-.6\,m/s)$$

$$Vy_f = 0.48\,m/s$$

## HORIZONTAL FORCE AND HORIZONTAL IMPULSE IN WALKING

The horizontal force and horizontal velocity for walking are shown in Figure 9.4. Horizontal force in walking typically begins with a brief propulsion phase, following by braking and propulsion.

### Example 3

Use the average force and formula to compute the horizontal velocity (Vx) for each phase of the horizontal force curve for the walking curve in Figure 9.4. The solution for the first propulsion phase from t = 0.0 to t = 0.04 s is shown below.

$$\bar{F}(\Delta t) = mV_f - mV_i$$

$$\bar{R}x(\Delta t) = mVx_f - mVx_i$$

$$26.1N(.04s) = 61kg(Vx_f) - 61kg(1.8\,m/s)$$

$$Vx_f = 1.82\,m/s$$

The solution for the braking phase from t = 0.04 to t = 0.4 s is shown below.

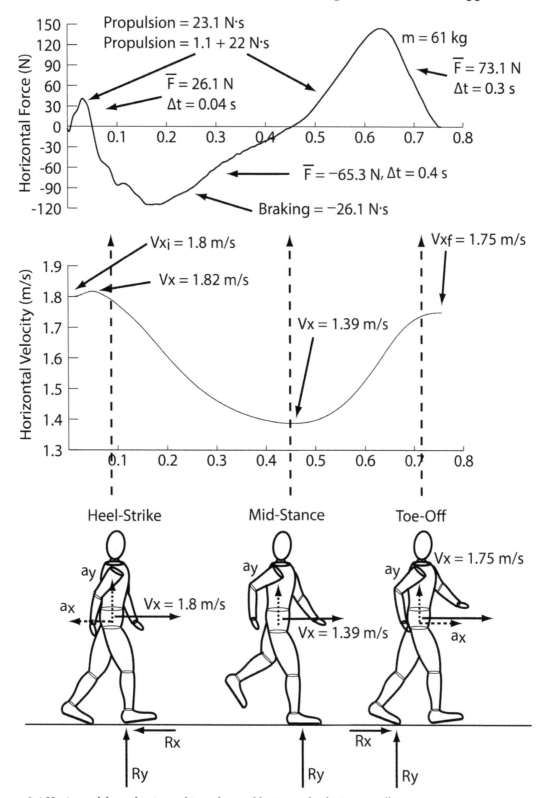

**Figure 9.4** Horizontal force, horizontal impulse, and horizontal velocity in walking.

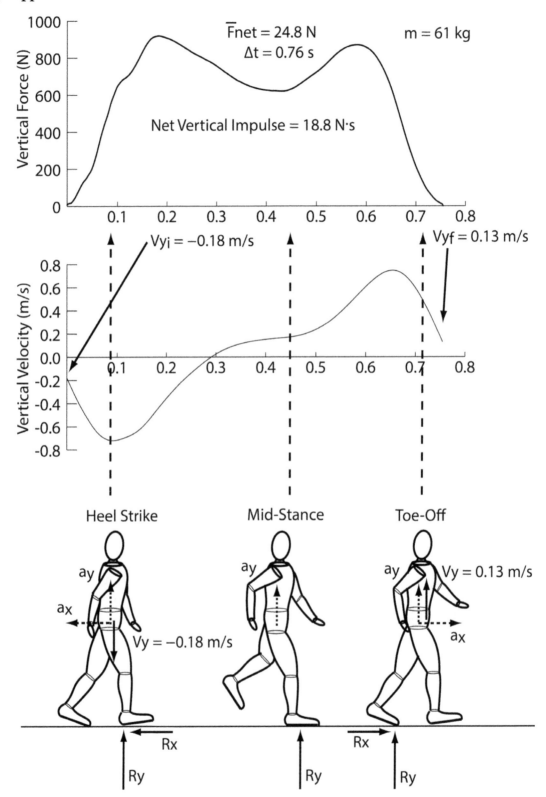

**Figure 9.5** Vertical force, vertical impulse and vertical velocity in walking.

$$\bar{F}(\Delta t) = mV_f - mV_i$$

$$\bar{R}x(\Delta t) = mVx_f - mVx_i$$

$$-65.3N(.4s) = 61kg(Vx_f) - 61kg(1.82\,m/s)$$

$$Vx_f = 1.39\,m/s$$

The solution for the second propulsion phase from t = 0.4 to t = 0.7 s is shown below.

$$\bar{F}(\Delta t) = mV_f - mV_i$$

$$\bar{R}x(\Delta t) = mVx_f - mVx_i$$

$$73.1N(.3s) = 61kg(Vx_f) - 61kg(1.39\,m/s)$$

$$Vx_f = 1.75\,m/s$$

## VERTICAL FORCE AND VERTICAL IMPULSE IN WALKING

Figure 9.5 shows the changes in vertical force and vertical velocity that occur during ground contact in walking. In addition, the free-body diagram is shown for three phases of the motion: heel-strike, mid-stance, and toe-off. The vertical force (Ry) is positive throughout ground contact. At heel-strike, the vertical velocity of the center of mass (Vy) is negative. Since the force Ry is opposing the walker's motion, it will cause the vertical velocity of the walker's center of mass to change from negative ($Vy_i = -.18$ m/s) at heel-strike to positive ($Vy_f = .13$ m/s) at toe-off.

### Example 4

Use the area under the curve impulse-momentum formula to compute the $Vy_f$ for the walking vertical force shown in Figure 9.5.

$$\int_{t_0}^{t_1} F\,dt = mV_f - mV_i$$

$$\int_0^{.76} (Ry + mg)\,dt = 61kg(Vy_f) - 61kg(-.18\,m/s)$$

$$18.8N \cdot s = 61kg(Vy_f) - 61kg(-.18\,m/s)$$

$$Vy_f = 0.13\,m/s$$

## IMPULSE-MOMENTUM IN THE VERTICAL JUMP

The vertical force and vertical velocity for a vertical jump are shown in Figure 9.6. At the start of the jump, the vertical velocity is Vy = 0 m/s. At approximately 0.26 s, the vertical velocity attains the greatest downward value of −0.8 m/s. The jumper stops moving downward at t = .42 s and finally at takeoff, the vertical velocity is 2.06 m/s. Use the area under the force–time formula to compute the vertical velocity at takeoff as follows:

$$\int_{t_0}^{t_1} F\,dt = mV_f - mV_i$$

$$\int_0^{.678} (Ry + mg)\,dt = 69kg(Vy_f) - 69kg(0\,m/s)$$

$$142N \cdot s = 69kg(Vy_f) - 69kg(0\,m/s)$$

$$Vy_f = 2.06\,m/s$$

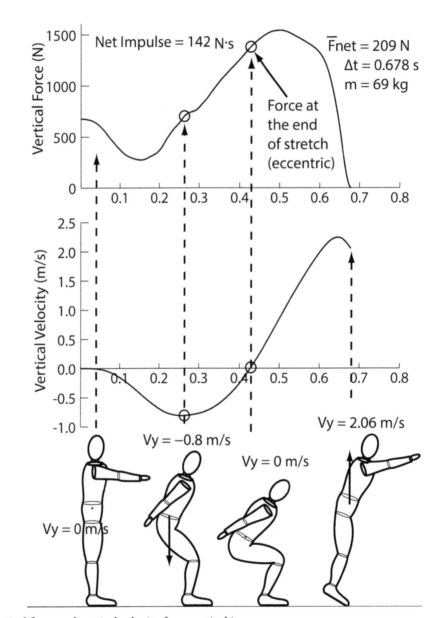

**Figure 9.6** Vertical force and vertical velocity for a vertical jump.

The dashed vertical line and circle at t = .26 s, on the graphs in Figure 9.6 indicate the time point in the motion when the jumper is moving downward the fastest. At this point, the jumper's center of mass has a vertical velocity of −0.8 m/s and the vertical force is equal to the jumper's body weight. The dashed vertical line and circle at t = .42 s on the velocity and force curves indicate the point in the motion when the jumper has stopped moving down and is about to begin moving upward. From the perspective of muscle mechanics, this point is the end of the eccentric loading of the knee extensors and the hip extensors.

### RELATIONSHIP BETWEEN FORCE AND ACCELERATION

Figure 9.7 shows the vertical force and vertical acceleration for a vertical jump. Examine the shape of the force and acceleration curves. They have exactly the same shape, they only differ by the units, the force has units of N and the acceleration has units of m/s². Recall from Newton's second law of motion (**law of acceleration**), that the acceleration is directly related to the force, and that it always occurs in the direction of the net force. Therefore, if you divide the force by mass you get acceleration. Since mass is a constant (m = 60 kg), dividing by mass

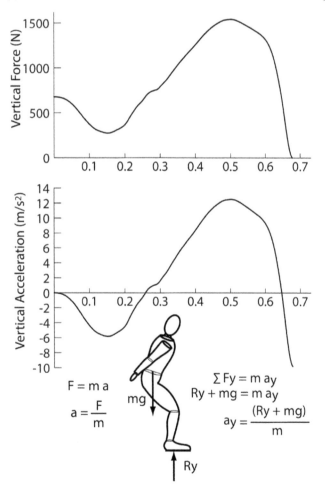

**Figure 9.7** Relationship between vertical force and vertical acceleration in a vertical jump.

does not change the shape of the curve. To compute the vertical acceleration from the force curve in Figure 9.7 just subtract body weight (mg) from the force, and then divide by mass. Thus, anytime you see a force curve you should intuitively know what the shape of acceleration curve would look like.

## SPEEDING UP, SLOWING DOWN, AND CONSTANT VELOCITY IN GAIT

Figure 9.8 shows examples of horizontal force curves with braking and propulsion impulses and the change in horizontal velocity for: speeding up, constant velocity, and slowing down. When running or walking at a constant velocity, the area of braking is equal to the area of propulsion. According to the impulse–momentum relationship, the change in horizontal

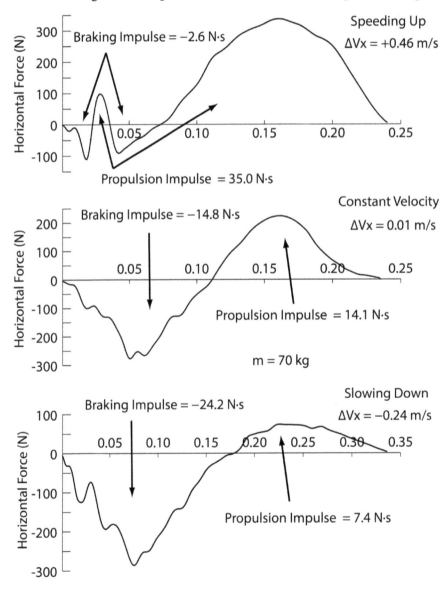

**Figure 9.8** Braking and propulsion impulses for speeding up, constant velocity, and slowing down, respectively.

velocity will be zero or close to zero (see the middle graph in Figure 9.8). When speeding up (accelerating) during walking or running, the area of propulsion will be greater than the area of braking and the change in velocity will be a positive number. The top graph in Figure 9.8 shows that the sum of the braking impulses was −2.6 N·s and the sum of the propulsion impulses was 35.0 N·s. Since the area of propulsion is much larger than the area of braking, the horizontal velocity (Vx) will change by +0.46 m/s for the runner after this ground contact. Finally, when slowing down (decelerating) the area of braking will be greater than the area of propulsion and the change in horizontal velocity will be negative. As shown in the bottom graph of Figure 9.8, the area of braking (−24.2 N·s) is greater than the area of propulsion (+7.4 N·s). Since the area of braking is greater than the area of propulsion, the runner's horizontal velocity will decrease by 0.24 m/s after this ground contact.

## COMPARISON BETWEEN HORIZONTAL AND VERTICAL FORCES

Figure 9.9 shows both horizontal and horizontal forces for walking and running. The magnitude of the forces for running is greater than the magnitude for walking. The time of ground contact is much longer for walking than running. The peak forces are greater in running than in walking.

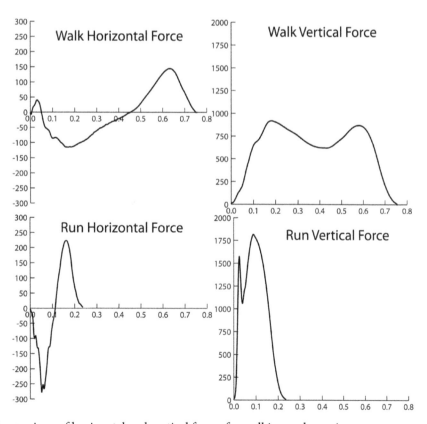

**Figure 9.9** Comparison of horizontal and vertical forces for walking and running.

## COMPUTING ACCELERATION AND VELOCITY FROM FORCE IN EXCEL

Figure 9.10 shows the first few rows of the Excel file "**Lesson 9 Practice.xls**", which is available from the course website. The data values in columns A, B, & C were collected in a biomechanics lab using a force plate of an individual walking over the force plate. Column B contains the horizontal reaction force (Rx) and column C contains the vertical reaction force (Ry). Cell F2 contains the walker's initial horizontal velocity at heel contact and cell G2 contains the initial vertical velocity of the center of mass at heel contact. Finally, cell L2 contains the subject's mass. Follow the instructions below to compute $a_x$, $a_y$, $v_x$, $v_y$ and the impulses using the force–time values.

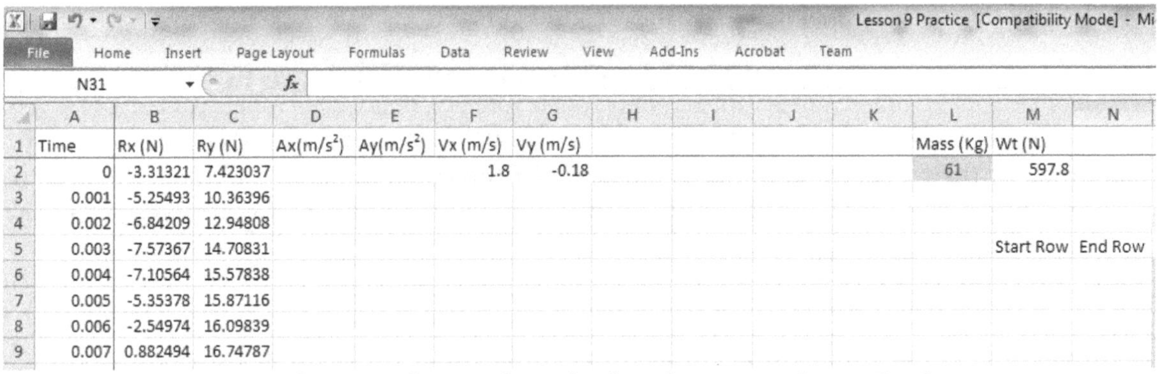

| | A | B | C | D | E | F | G | H | I | J | K | L | M | N |
|---|---|---|---|---|---|---|---|---|---|---|---|---|---|---|
| 1 | Time | Rx (N) | Ry (N) | Ax(m/s²) | Ay(m/s²) | Vx (m/s) | Vy (m/s) | | | | | Mass (Kg) | Wt (N) | |
| 2 | 0 | -3.31321 | 7.423037 | | | 1.8 | -0.18 | | | | | 61 | 597.8 | |
| 3 | 0.001 | -5.25493 | 10.36396 | | | | | | | | | | | |
| 4 | 0.002 | -6.84209 | 12.94808 | | | | | | | | | | | |
| 5 | 0.003 | -7.57367 | 14.70831 | | | | | | | | | | Start Row | End Row |
| 6 | 0.004 | -7.10564 | 15.57838 | | | | | | | | | | | |
| 7 | 0.005 | -5.35378 | 15.87116 | | | | | | | | | | | |
| 8 | 0.006 | -2.54974 | 16.09839 | | | | | | | | | | | |
| 9 | 0.007 | 0.882494 | 16.74787 | | | | | | | | | | | |

**Figure 9.10** Computing acceleration, velocity and impulse from force-time values in Excel .

### Horizontal Acceleration

To compute horizontal acceleration ($a_x$) enter the following equation in cell D2 and copy it down the rest of the column to row 757, where the data ends:

=B2/$L$2

See Figure 9.11 for an explanation of how the equation above is derived.

Compute ax

$$\Sigma Fx = m\, ax$$
$$Rx = m\, ay$$
$$ax = \frac{Rx}{m}$$
$$=B2/\$L\$2$$

**Figure 9.11** Use F = ma to compute horizontal acceleration from force and mass.

### Vertical Acceleration

To compute vertical acceleration ($a_y$) enter the following equation in cell E2 and copy it down the rest of the column to row 757, where the data ends:

=(C2+($L$2*-9.8))/$L$2

Compute ay

$$\Sigma Fy = m\, ay$$
$$Ry + mg = m\, ay$$
$$ay = \frac{Ry + mg}{m}$$
$$=(C2+(\$L\$2*-9.8))/\$L\$2$$

**Figure 9.12** Use F = ma to compute vertical acceleration and velocity from force and mass.

See Figure 9.12 for an explanation of how the equation above is derived.

## Horizontal Velocity

To compute horizontal velocity (Vx), enter the following equation in cell F3 and copy it down the rest of the column to row 757, where the data ends:

=((B3*0.001)+($L$2*F2))/$L$2

Compute Vx

$$R_x(t) = mV_{xf} - mV_{xi}$$

$$V_{xf} = \frac{R_x(t) + mV_{xi}}{m}$$

=((B3*0.001)+($L$2*F2))/$L$2

**Figure 9.13** Use impulse-momentum to compute horizontal velocity from force, mass, and time.

See Figure 9.13 for an explanation of how the equation above is derived.

## Vertical Velocity

To compute vertical velocity (Vy), enter the following equation in cell G3 and copy it down the rest of the column to row 757, where the data ends:

= (((C3 + ($L$2*-9.8))*0.001) + ($L$2*G2)) / $L$2

Compute Vy

$$(R_y + mg)(t) = mV_{yf} - mV_{yi}$$

$$V_{yf} = \frac{(R_y + mg)t + mV_{yi}}{m}$$

= (((C3 + ($L$2*-9.8))*0.001) + ($L$2*G2)) / $L$2

**Figure 9.14** Use impulse-momentum to compute vertical velocity from force, mass, and time.

See Figure 9.14 for an explanation of how the equation above is derived.

## Computing Impulses in Excel

Figure 9.15 shows the horizontal force curve and the "Horizontal Impulse-Momentum" section of the Excel spreadsheet. After making a graph of horizontal force, you must enter the start and end row numbers where the force curve crosses the Y = 0 axis. These row numbers should be placed in columns "M" and "N". The graph in Figure 9.15 shows the corresponding rows where the force curve crosses the Y = 0 axis. The horizontal force begins with a negative value in row 2 of column B and the last negative value for Rx in column B is in row 8. To get the impulse of this area, enter the numbers "2" and "8" in cells M6 and N6, respectively. The horizontal force is positive in row 9 of column B and the last positive value for Rx in column B is in row 51. To get the impulse of this area, enter the numbers "9" and "51" in cells M7 and N7, respectively. Complete the analysis by entering the start and end rows of all positive and negative phases of the horizontal force (Rx).

To find these row numbers, look in Column B. As shown in the Figure 9.15, the force is negative from rows 2 – 8 and row 9 is the start of the positive phase. Enter "2" in cell M6 and "8" in cell N6 to indicate the start and end of the first negative phase. Then scroll down to find the end of the positive phase that begins in row 9. Enter "9" in cell M7 and "51" in cell N7

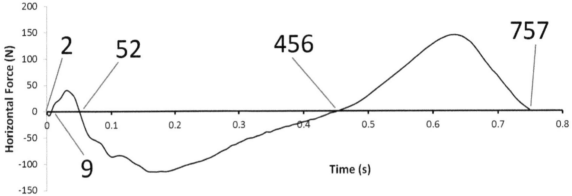

| L | M | N | O | P | Q | R | S | T | U |
|---|---|---|---|---|---|---|---|---|---|
| Mass (Kg) | Wt (N) | | | | | | | | |
| 61 | 597.8 | | | | | | | | |
| | | | | Horizontal Impulse - Momentum Variables | | | | F Ave | Change V |
| | Start Row | End Row | Vx i | Vx f | Change Vx | dt | F Ave | Impulse | Impluse |
| | 2 | 8 | 1.80 | 1.80 | 0.00 | 0.01 | -5.43 | -0.03 | -0.03 |
| | 9 | 51 | 1.80 | 1.82 | 0.02 | 0.04 | 25.31 | 1.06 | 1.09 |
| | 52 | 455 | 1.82 | 1.39 | -0.43 | 0.40 | -64.87 | -26.14 | -26.21 |
| | 456 | 757 | 1.39 | 1.75 | 0.36 | 0.30 | 72.89 | 21.94 | 22.01 |

**Figure 9.15** To compute the impulse of each positive and negative horizontal force phase, determine the start and end of each positive and negative phase of the horizontal force. In the example above, the horizontal force is negative from row 2 to row 8, the force is positive from row 9 to row 51, negative from rows 52-455, and positive from rows 456-757. The Excel spreadsheet has macro formulas that compute the change in horizontal velocity, the average force, and the impulse for each start and end row combination entered in columns M and N.

to indicate the start and end of the first positive phase. Then scroll down to find the end of the negative phase that begins in row 52 (Figure 9.15).

Enter "52" in cell M8 and "455" in cell N8 to indicate the start and end of the first negative phase. Then, scroll down to find the end of the positive phase that begins in row 456. Finally, enter "456" in cell M9 and "757" in cell N9 to indicate the start and end of the first negative phase. Then scroll down to find the end of the positive phase that begins in row 9 (see Figure 9.19).

The horizontal and vertical impulse-momentum variable sections are designed to automatically compute impulses using both the change in momentum and the average force times Δt. Once you enter the start and end rows, it computes the velocity for the start and endpoints, the time between the two points, the average force, and the impulse using both methods. The horizontal and vertical impulse sections are programmed to compute the impulse variables for up to 8 different regions on a curve. The computed horizontal and vertical impulses for each phase are shown in Figure 9.15.

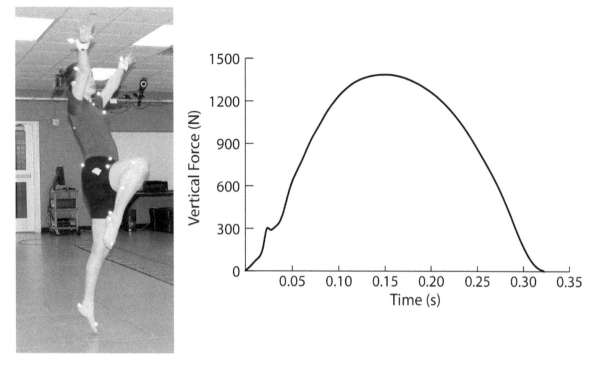

**Figure 9.16** Vertical force applied by a dancer to perform a leap.

## COMPUTING IMPULSE USING THE CHANGE IN MOMENTUM

Impulse can be computed using the right-hand side of the impulse momentum equation. When using the right-hand side of the impulse momentum equation to compute impulse, you need an initial and final velocity and the mass of the object or performer. Velocity before and after a force application can be determined using a video camera or a cell phone camera. Begin by filming the object or performer before and after the force application. Then scale the data to life-size dimensions. After scaling the data, compute the velocity immediately before the force and immediately after the force. Figure 9.16 shows a 58 kg dancer taking off for a leap and the vertical reaction force she imparted during the ground contact. Prior to the force application, she had an initial vertical velocity of 0 m/s, and after the force was applied her final vertical velocity was 1.46 m/s. The net impulse she applied to perform the leap can be computed as follows:

$$(\bar{R}y + mg)(\Delta t) = mVy_f - mVy_i$$
$$\text{Impulse} = mVy_f - mVy_i$$
$$\text{Impulse} = (58\,\text{kg})(1.46\,\text{m/s}) - (58\,\text{kg})(0\,\text{m/s})$$
$$\text{Impulse} = 84.68\,\text{N}\cdot s$$

The dancer applies a vertical impulse of 84.68 N·s, which changes her vertical velocity from 0 m/s to 1.46 m/s at takeoff. Notice that impulse was computed using only the change in velocity before and after the force was applied.

## SUMMARY

This chapter demonstrates the relationship between force, the impulse it causes, and the change in velocity produced by the impulse and force over some time interval. In addition, this chapter illustrates how acceleration, velocity and impulses can be computed from force-time data.

## REVIEW QUESTIONS

1. State both the average force and the integral versions of the impulse–momentum equation. For each equation, explain how you could compute impulse using both the right and left hand side of the equation.
2. Describe the relationship between braking and propulsion impulses in running and the horizontal velocity of the runner's center of mass.
3. Describe the steps necessary and the equations needed to compute horizontal acceleration and horizontal velocity from a horizontal force–time curve in running.
4. Describe the steps necessary and the equations needed to compute vertical acceleration and vertical velocity from a vertical force–time curve in a vertical jump.
5. Use the horizontal force curve in Figure 9.17 to answer questions a-f.
   a. Draw an xy graph of what you think the horizontal acceleration would look like.

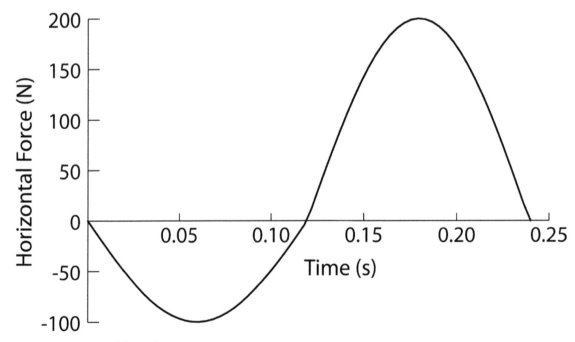

Figure 9.17 Horizontal force during running.

b.  Draw an xy graph of what you think the horizontal velocity curve would look like. Assume that the initial horizontal velocity is 4 m/s.
c.  Which impulse is greater, braking or propulsion?
d.  At toe-off, will the runner's horizontal velocity be greater or less than the initial horizontal velocity of 4 m/s at heel-strike?

## Suggested Supplemental Resources

Cutnell, J. D. and K. W. Johnson (2008). Physics. Hoboken, John Wiley & Sons.

Enoka, R. M. (2008). Neuromechanics of Human Movement. Champaign, Human Kinetics.

Griffiths, I. W. (2006). Principles of Biomechanics & Motion Analysis. Philadelphia, Lippincott Williams & Wilkins.

Hall, S. J. (2007). Basic Biomechanics. Boston, McGraw Hill.

Hamill, J. and K. M. Knutzen (2009). Biomechanical Basis of Human Movement. Philadelphia, Lippincott Williams & Wilkins.

Hibbeler, R. C. (2010). Engineering Mechanics: Dynamics. Upper Saddle River, Prentice Hall.

Hibbeler, R. C. (2010). Engineering Mechanics: Statics. Upper Saddle River, Prentice Hall.

McGinnis, P. M. (2013). Biomechanics of Sport and Exercise. Champaign, Human Kinetics.

Winter, D. A. (2009). Biomechanics and Motor Control of Human Movement. Hoboken, John Wiley & Sons, Inc.

# Chapter 10

## WORK–ENERGY–POWER

### STUDENT LEARNING OUTCOMES

**After reading Chapter 10, you should be able to:**
- Define work, energy, and power.
- Define the **work–energy principle**.
- Define the **law of conservation of energy**.
- Demonstrate how to solve work, energy, and power problems.

### WORK

Work is defined as the magnitude of force times distance. Work is a scalar. The units for work are N·m or joules (J). Work for a constant or average force is computed as follows:

$$W = F(y_f - y_i)$$

where W is work in N·m, F is force, $y_f$ is final position, and $y_i$ is initial position. Work can be either positive or negative, depending upon whether the force acts in the same direction as the displacement, or the opposite direction of the displacement. The jumper shown in Figure 10.1, does positive work during the takeoff for a vertical jump. In the takeoff phase of the jump, the work is positive because the displacement and force act in the same direction. During the

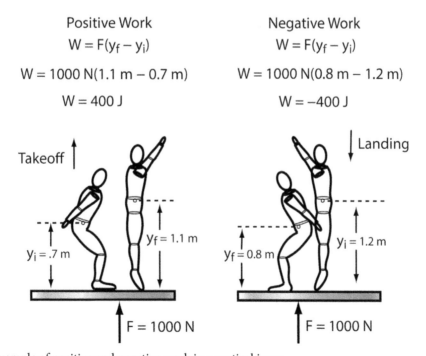

Positive Work
$W = F(y_f - y_i)$
$W = 1000\ N(1.1\ m - 0.7\ m)$
$W = 400\ J$

Negative Work
$W = F(y_f - y_i)$
$W = 1000\ N(0.8\ m - 1.2\ m)$
$W = -400\ J$

Takeoff ↑     $y_f = 1.1\ m$     $y_i = .7\ m$     F = 1000 N

↓ Landing     $y_i = 1.2\ m$     $y_f = 0.8\ m$     F = 1000 N

**Figure 10.1** Example of positive and negative work in a vertical jump.

landing phase of the motion, the jumper does negative work since the force acts in the opposite direction of the motion.

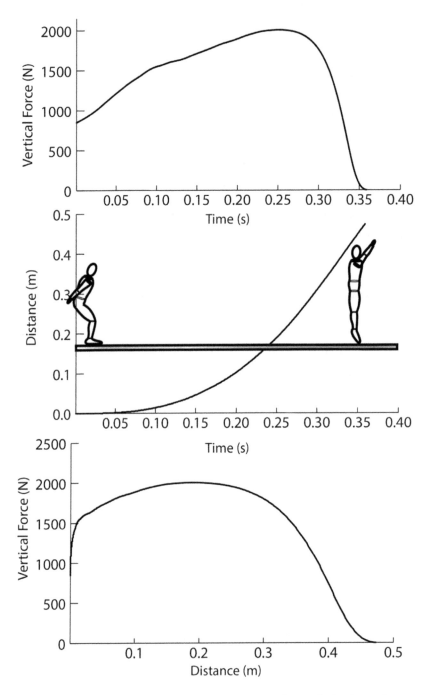

**Figure 10.2** The work done by a variable force is equal to the area underneath the force-distance graph. The top graph shows vertical force (N), the middle graph show distance (m), the bottom graph shows force-distance. The area underneath the force-distance graph is equal to the work done in joules.

No work is done when pushing on an immovable object, such as a wall, since the displacement is zero. In addition, no work is done while holding a weight, since the displacement is zero. As a result, isometric contractions do no mechanical work.

## WORK DONE BY A VARIABLE FORCE

The work done by a variable force is equal to the area under the force–distance graph. Figure 10.2 below shows the vertical force for a vertical jump and the distance over which the center of mass is raised. The bottom plot depicts the force–distance graph. The area under the force–distance graph can be computed as follows:

$$W = \int_{y_0}^{y_1} F \times \Delta y$$

where W is work, F is force, and $\Delta y$ is the change in distance.

## KINETIC AND POTENTIAL ENERGY

Kinetic energy is the energy a body possesses due to its motion. Like work, kinetic energy is a scalar. The units for kinetic energy are joules (J). Kinetic energy is computed as follows:

$$KE = \frac{1}{2}mVy^2$$

where KE is kinetic energy in J, m is mass in kg, and Vy is the vertical velocity.

Potential energy, sometimes referred to as gravitational potential energy, is the energy a body possesses due to its position. The units for potential energy are J. Potential energy is a scalar. Potential energy is computed as follows:

$$PE = mgh$$

where PE is potential energy, m is mass in kg, g is acceleration due to gravity ($+9.8$ m/s$^2$) and, h is height in m. Note that the "**g**" used in the computation of potential energy is **+9.8 m/s$^2$**.

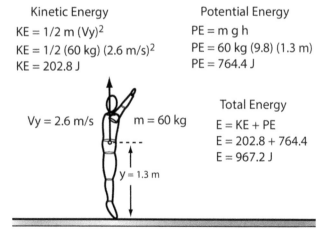

Figure 10.3 Example of kinetic, potential, and total energy in a vertical jump.

## TOTAL MECHANICAL ENERGY

The total mechanical energy a body possesses is simply the sum of kinetic and potential energy. Total energy is a scalar, with units of J. Figure 10.3 shows the kinetic, potential, and total energy for a vertical jump at takeoff. The total mechanical energy is the sum of the kinetic and potential energy as follows:

$$E = KE + PE$$

where E is total mechanical energy in J, KE is kinetic energy, and PE is potential energy.

## WORK-ENERGY PRINCIPLE

According to the **work–energy principle** (sometimes called the **work–energy theorem**), the work done is equal to the change in energy. When a net force does work on an object, the force changes the energy of the object. The **work–energy principle** applies whenever you are in contact with the ground or some external object. As long as you can apply a force over some distance, you can change the energy of an object or performer. **Work–energy principle** is computed using the following equation:

$$W = \frac{1}{2}m(Vy_f)^2 - \frac{1}{2}m(Vy_i)^2 + mgh_{y_f - y_i}$$

W = 1/2 m (Vy$_f$)$^2$ – 1/2 m (Vy$_i$)$^2$ + m g (y$_f$ – y$_i$)

W = 1/2 (86.73) (2.67)$^2$ – 1/2 (86.73) (0)$^2$ + (86.73)(+9.8)(1.373 – 0.9)

W = 711.17 J

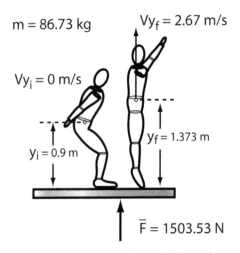

m = 86.73 kg        Vy$_f$ = 2.67 m/s

Vy$_i$ = 0 m/s

y$_f$ = 1.373 m

y$_i$ = 0.9 m

F̄ = 1503.53 N

**Figure 10.4** Using the **work-energy principle** to compute the work done during the pushing phase of a vertical jump. The 711.17 J of positive work changes the jumper's vertical velocity from 0 m/s to 2.67 m/s at takeoff.

where W is work, m is mass, Vy is vertical velocity, g is +9.8 m/s², and $h_{yf-yi}$ is the change in height. Figure 10.4 gives an example of using the **work–energy principle** to calculate the work done between two different phases of a vertical jump. Initially, the jumper has a height of 0.9 m and a vertical velocity of 0 m/s. The jumper then applies a force over a distance of .473 m. The force does work on the jumper and increases the energy of the jumper. The 711.17 J of work done is equal to the change in energy between the initial and final position in Figure 10.4.

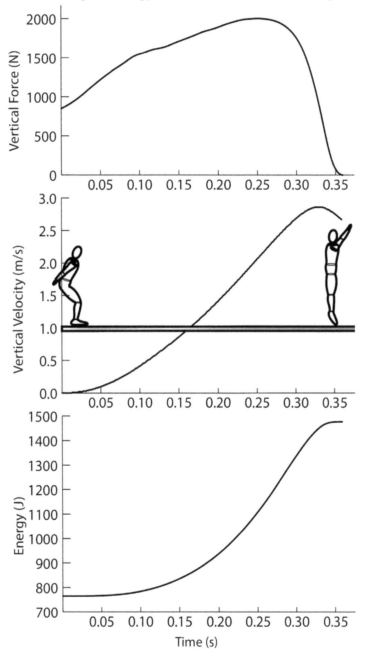

**Figure 10.5** Example of the **work-energy principle.** Since gravity is not the only force acting on the system, the work done is equal to the change in energy (top vertical force, middle vertical velocity, bottom energy).

Figure 10.5 shows the change in energy as a function of time for the vertical jump problem in Figure 10.4. The **work–energy principle** applies whenever the performer is in contact with the ground or a fixed object. While in contact with the ground or a fixed object, the performer is able to apply a force over some distance, which then changes the total energy. The energy–time curve at the bottom of Figure 10.5 illustrates the change in energy caused by the work done in the jumping motion. When gravity is not the only force operating on a system, the energy of the system can be changed according to the **work–energy principle**. The next section will describe the **law of conservation of energy**. Unlike the **work–energy principle**, where the change in energy is equal to the work done, the in the air the total energy is constant since gravity is the only force acting on the system.

## Law of Conservation of Energy

According to the law of conservation of energy, when gravity is the only force operating, the total energy of a system is constant. As a result, whenever an object or performer is in the air, the total energy is constant. Figure 10.6 gives an example of the **law of conservation of energy** for a diver. At takeoff, the diver has 1250 J of energy. The total energy is constant throughout the motion. On the way up, the diver's vertical velocity decreases, which causes

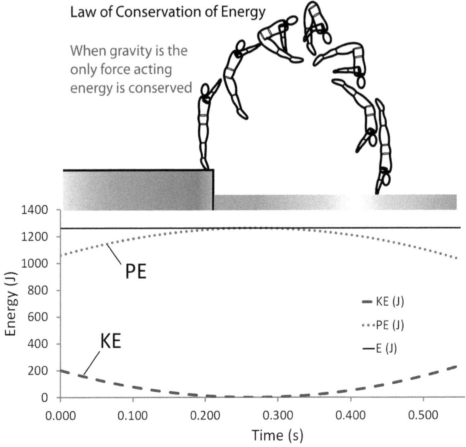

**Figure 10.6** Law of conservation of energy. Energy is constant when gravity is the only force acting on a system.

kinetic energy to decrease and the diver's height increases, which causes potential energy to increase. At the peak, the vertical velocity is zero, so kinetic energy is zero. At the peak, the total energy is equal to the potential energy. On the way down, the diver loses potential energy and gains kinetic energy. At any point in the flight, the total energy is equal to the sum of kinetic and potential energy.

## WORK-ENERGY PRINCIPLE OR LAW OF CONSERVATION OF ENERGY?

The **work-energy principle** applies to any situation where the performer or object can apply forces which can do work and change the energy of the object or performer. When the performer is on the ground, energy is not conserved and as a result, the performer can change his/her energy. In the air, energy is conserved and as a result, the total energy of a system is

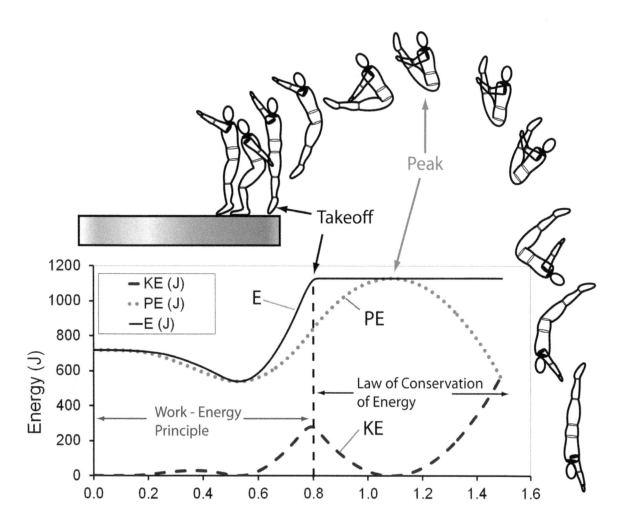

**Figure 10.8** Comparison of the **work-energy principle** to the **law of conservation of energy** in diving. From t = 0 s to t = 0.8 s, the diver is in contact with the platform where he apply forces against the ground and does work. From t = 0.8 s to t = 1.6 s, the diver is in the air, so total energy (E) is constant.

constant. Figure 10.8 depicts the differences between the **work–energy principle** and the **law of conservation of energy**. On the left-hand side the diver applies forces against the ground, which changes the diver's energy. Once in the air, the diver's energy is constant.

## Finding Peak Height using the Law of Conservation of Energy

When gravity is the only force, the total energy of a system is constant. At the peak, the vertical velocity is zero and all total energy is equal to the potential energy. As a result, the peak height obtained can be obtained using the following formula:

$$h = \frac{E}{mg}$$

where h is peak height, E is the total energy, m is mass, and g is +9.8 m/s$^2$.

## Peak Height Sample 1

In Figure 10.6 the diver has a total energy (E) of 1250 J and a mass of 60 kg. Compute the peak height obtained by the diver. The computation of the peak height obtained by the diver in Figure 10.6 is shown below. The diver reaches a peak height of 2.13 m.

$$h = \frac{E}{mg}$$

$$h = \frac{1250J}{60kg(+9.8)}$$

$$h = 2.13m$$

**Figure 10.7** Find peak height of a dance leap using the **law of conservation of energy**.

## Peak Height Sample 2

The 58 kg dancer shown in Figure 10.7 takes off with a vertical velocity (Vy) of 1.46 m/s and a height (h) of 1.13 m. Find peak height using energy.

$$KE = \frac{1}{2}mVy^2$$

$$KE = \frac{1}{2}(58kg)(1.46m/s)^2$$

$$KE = 61.82J$$

$$PE = mgh$$

$$PE = (58kg)(+9.8)(1.13m)$$

$$PE = 642.29J$$

$$h = \frac{E}{mg}$$

$$E = KE + PE$$

$$E = 61.82J + 642.29J$$

$$E = 704.11J$$

$$h = \frac{704.11J}{(58kg)(+9.8)}$$

$$h = 1.24m$$

As shown above, to find the peak height using energy, begin by finding kinetic, potential, and total energy. Then divide the total energy by mass times g to get peak height. The dancer in Figure 10.7 reaches a peak height of 1.24 m.

## POWER

Power is an important concept in human movement. We often refer to an athlete as being powerful or explosive. Athletes who train to improve power typically use resistance loads of 50–80% of their 1 RM. When training to improve power, it is essential that each repetition be performed as rapidly as possible, so thus on each attempt, the individual lifts the weight at his or her maximal velocity throughout the range of motion. Performing each lift at maximal velocity imposes a high force overload on the muscles while simultaneously training the central nervous system to learn a neural drive that optimizes muscle production.

**Power** is defined as the rate of doing work. Power is also the rate of energy transfer. Power incorporates both work and time, or force and velocity. The units for power are watts (W). Like work and energy, power is a scalar. Power can be computed using either of the formulas below:

$$P = \frac{Work}{Time}$$

Power = Force ×Velocity

where work is in W and time is in s, or force is in N and velocity is in m/s. Figure 10.8 shows force, velocity, and power for a vertical jump. The vertical force–time data were recorded using a force plate. The vertical velocity–time curve was computed from the force–time data using impulse–momentum.

$$\int_{t_0}^{t_1} (Ry + mg)\, dt = mVy_f - mVy_i$$

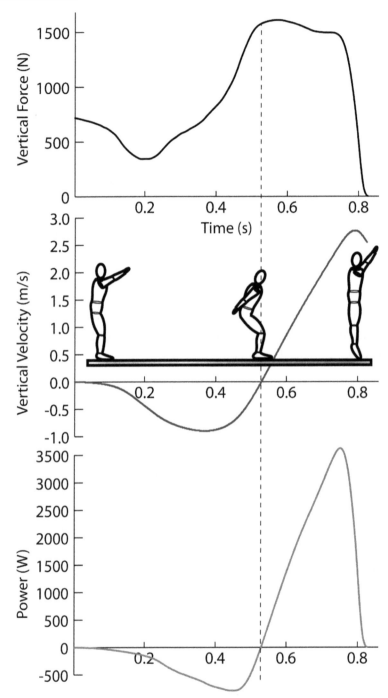

**Figure 10.8** Power is force × velocity, or work divided by time. Multiplying the vertical force (top graph) by the vertical velocity (middle graph) yields the power-time curve (bottom graph).

After computing vertical velocity as a function of time, power was computed by multiplying the vertical force times vertical velocity.

$$Power = Force \times Velocity$$

In the power–time curve shown in Figure 10.8, the power is negative to the left of the dashed vertical line (0.53 s). The negative power from 0–0.53 s can be attributed to eccentric actions of the hip, knee, and ankle extensors. The positive power from t = 0.53 s to takeoff at t = 0.82 s can be attributed to the concentric force production of the hip, knee, and ankle extensors. The power–time curve in Figure 10.8 can also be obtained using the **work–energy principle** to compute the work done as a function of time.

## ROTATIONAL KINETIC ENERGY

A rotating body possesses kinetic energy due to its rotational motion. The rotational kinetic energy ($KE_R$) of a rigid object rotating with an angular velocity ($\omega$) about some axis of rotation with a moment of inertia ($I$) is defined as follows:

$$KE_R = \frac{1}{2}I\omega^2$$

where $KE_R$ is rotational kinetic energy in J, $I$ is the moment of inertia about the axis of rotation in kg·ms², and $\omega$ is the angular velocity in r/s. The angular velocity must be expressed in radians per second. Rotational kinetic energy is a scalar. The moment of inertia represents the object's resistance to angular change about the axis. Moment of inertia is a function of mass and how the mass is distributed about the axis of rotation. Figure 10.9 shows a diver in the air rotating with an angular velocity of 1.3 r/s and a moment of inertia of 12 kg·ms². The rotational kinetic energy for the diver is computed as follows:

$$KE_R = \frac{1}{2}I\omega^2$$

$$KE_R = \frac{1}{2}(12kg \cdot m^2)(1.3r/s)^2$$

$$KE_R = 10.14J$$

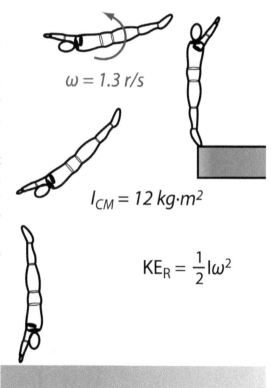

Figure 10.9 Rotational kinetic energy in diving.

## ROTATIONAL WORK

The rotational work done by a constant torque rotating an object some angular distance is computed using the following equation:

$$W_R = \tau(\theta_f - \theta_i)$$

where $W_R$ is rotational work in watts, $\tau$ is the torque in N·m, and $\theta_f$ is the final angle, and $\theta_i$ is the initial angle. Rotational work is a scalar. Rotational work is positive when the torque and the angular distance are in the same direction and negative when they are in opposite directions. The angles must be expressed in radians.

### Sample Rotational Work Problem

A weight lifter applies a constant torque ($\tau$) of 50 N·m while doing an elbow curl (see Figure 10.10). The lift begins at an initial elbow angle of 0.44 r and finishes at a final angle of 1.57 r. Determine the work done by the constant torque.

$$W_R = \tau(\theta_f - \theta_i)$$
$$W_R = 50N \cdot m(1.57r - 0.44r)$$
$$W_R = 56.5\,J$$

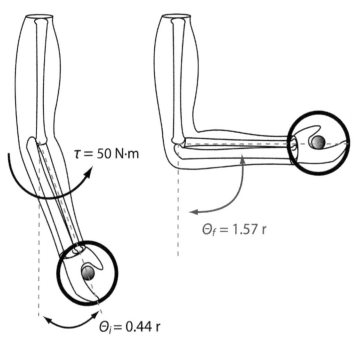

**Figure 10.10** Rotational work for a constant torque is computed by multiplying the torque (50 N·m) times the angular displacement. The positive 56.6 J of rotational work changes the elbow angle from 0.44 r to 1.57 r.

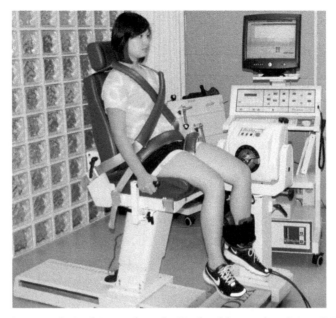

**Figure 10.11** An isokinetic strength machine such as the Biodex (shown above) or a Cybex can be used to measure isometric, concentric, and eccentric torque. Isokinetic machines can be configured to measure torque of ankle, knee, hip, elbow, and wrist joints.

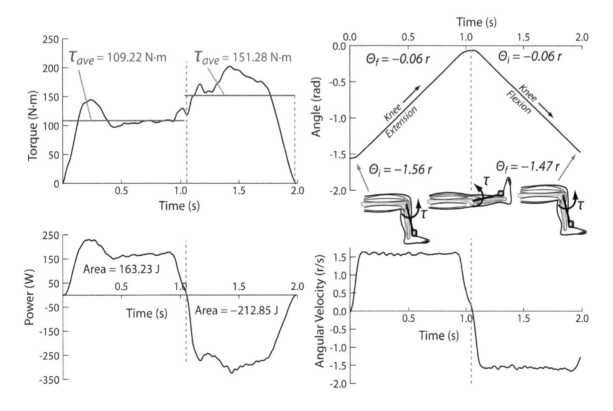

**Figure 10.12** Torque (upper left), angle (upper right), angular velocity (lower right) and power (lower left) for a concentric and eccentric isokinetic knee joint exercise.

## ROTATIONAL WORK FOR A NON-CONSTANT TORQUE

There are three different methods for computing the rotational work for a non-constant torque:

1. Determine the average torque and then multiply the average torque by the angular displacement.
2. Use integration to determine the area underneath the torque–angle graph.
3. Compute power from torque and angular velocity and then integrate the power curve with respect to time.

In humans, joint torque can be directly measured using an isokinetic strength machine (see Figure 10.11 for an example). An isokinetic machine is a specialized strength testing device that provides variable resistance at a constant (fixed) velocity. Isokinetic devices can measure joint torque concentrically and eccentrically at velocities ranging from 0–500 d/s. Once the velocity is set, the device provides variable resistance throughout the range of motion. The velocity of movement remains fixed, no matter how much effort is exerted by the subject. Isokinetic devices provide variable resistance equal and opposite to the force exerted by the subject, and as a result, subjects must be explicitly instructed to attempt to generate maximal torque throughout the range of motion being tested. Isokinetic devices can also be used to measure isometric strength. Isokinetic devices directly measure angle, angular velocity, and torque. After measuring torque, rotational work, power, and energy can then be computed. Isokinetic devices are typically used for strength testing and injury rehabilitation.

Figure 10.12 shows torque–time, angle–time, angular velocity–time and power–time curves for an isokinetic knee exercise. The angle–time graph in the upper right of Figure 10.12 shows the change in knee joint angle as a function of time. The left side of the angle–time curve is a concentric movement. During this concentric phase, the knee extensor muscles are generating an extension torque while the muscles are shortening. The right side of the angle–time curve is an eccentric movement. During the eccentric phase, the knee extensor muscles are generating an extension torque while the muscles are lengthening. The average torque for the concentric and eccentric phase is shown on the torque–time curve. The work done during the concentric phase can be computed as follows:

$$W_R = \tau_{ave}(\theta_f - \theta_i)$$
$$W_R = 109.22N \cdot m(-0.06r - -1.56r)$$
$$W_R = 163.83J$$

The work done during the eccentric phase is:

$$W_R = \tau_{ave}(\theta_f - \theta_i)$$
$$W_R = 151.28N \cdot m(-1.47r - -0.06r)$$
$$W_R = -213.30J$$

## ROTATIONAL POWER

Rotational power is the rate of doing angular work ($W_R$) or the rate of energy transfer. Power incorporates both angular work and time, or torque and angular velocity. The units for power are watts (W). Like angular work and rotational kinetic energy, rotational power is a scalar. Power can be computed using either of the formulas below:

$$P_R = \frac{W_R}{t}$$

where $P_R$ is rotational power in watts, $W_R$ is rotational work in J and t is time, or as follows:

$$P_R = \tau \times \omega$$

where $P_R$ is rotational power in W, $\tau$ is torque in N·m, and $\omega$ is the angular velocity in r/s. The angular velocity must be expressed in radians/second. The power curve shown in Figure 10.12 was computed by multiplying the torque by the angular velocity.

Rotational work can be computed by integrating a rotational power curve with respect to time. As shown in the power curve in Figure 10.11, the area of the concentric phase of the power–time curve is 163.83 J, which is the work done during the concentric phase. Finally, the area of the eccentric phase of the power–time curve in Figure 10.12 is –213.30 J, which is the work done during the eccentric phase.

## SUMMARY

This chapter illustrates how human movement can be analyzed using work, energy, and power. Both linear and angular components of work, energy, and power were explained.

## REVIEW EXERCISES

1. Define work and power.
2. How is kinetic, potential, and total mechanical energy calculated?
3. What is the difference between the work–energy principle and the law of conservation of energy? Under what conditions does each apply?
4. Define rotational work and describe how rotational work is calculated.
5. A 75 kg skate boarder takes off with a vertical velocity (Vy) of 2.9 m/s and a height ($y_i$) of 1.2 m. Find the peak height using energy.
6. A 50 kg dancer lands with an initial vertical velocity ($Vy_i$) of –2.8 m/s and an initial height of 1.22 m. The dancer applies a force against the ground to reduce her momentum. After 0.5 s. she has a final vertical velocity ($Vy_f$) 0.0 m/s and a final height ($y_f$) of 0.88 m. Compute the work done and average power.

## SUGGESTED SUPPLEMENTAL RESOURCES

Cutnell, J. D. and K. W. Johnson (2008). Physics. Hoboken, John Wiley & Sons.
Enoka, R. M. (2008). Neuromechanics of Human Movement. Champaign, Human Kinetics.

Griffiths, I. W. (2006). Principles of Biomechanics & Motion Analysis. Philadelphia, Lippincott Williams & Wilkins.

Hall, S. J. (2007). Basic Biomechanics. Boston, McGraw Hill.

Hamill, J. and K. M. Knutzen (2009). Biomechanical Basis of Human Movement. Philadelphia, Lippincott Williams & Wilkins.

Hibbeler, R. C. (2010). Engineering Mechanics: Dynamics. Upper Saddle River, Prentice Hall.

Hibbeler, R. C. (2010). Engineering Mechanics: Statics. Upper Saddle River, Prentice Hall.

McGinnis, P. M. (2013). Biomechanics of Sport and Exercise. Champaign, Human Kinetics.

Winter, D. A. (2009). Biomechanics and Motor Control of Human Movement. Hoboken, John Wiley & Sons, Inc.

# Chapter 11

## ANGULAR KINETICS – COMPUTING TORQUE

### STUDENT LEARNING OUTCOMES

**After reading Chapter 11, you should be able to:**
- Define torque.
- Define moment of inertia.
- Describe the steps in computing torque.
- Describe Newton's laws of angular motion.

### THE EFFECTS OF TORQUES ON RIGID OBJECTS

When a force is applied thru the center of mass of an object, the object will translate in the direction of the force. The magnitude of the translation is directly related to the magnitude of the net force, and inversely related to the object's resistance to linear change. In Figure 11.1, the force is applied directly thru the center of mass of the rigid object and, as a result, the object translates in the direction of the net force.

When a force is not applied directly thru the center of mass of a rigid object, it will cause both translation and rotation. In Figure 11.2, the net force does not act thru the center of mass of the rigid object and as a result, it causes both translation and rotation. A force that causes rotation is called a torque.

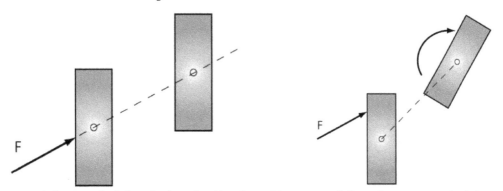

**Figure 11.1** A force applied directly thru the CM of an object causes the object to translate in the direction of the force.

**Figure 11.2** A force that is not applied thru the CM of the object causes translation and rotation.

### COMPUTING TORQUE

**Torque** is defined as a force that causes rotation, or tends to cause, rotation about some axis or point. Torque is represented by the symbol $\tau$ (Greek letter "tau"). The torque caused by a force is equal to the magnitude of the force times the lever arm. Torque is also referred to as the **moment of force**. Torque is computed using the following equation:

$$\tau = \pm |F| \times l$$

where $\tau$ is torque in N·m, $|F|$ is the magnitude of the force in N, and $l$ is the lever arm in m. Torque is a vector defined by the right-hand rule. A torque is positive when the force causes, or tends to cause a counterclockwise (CCW) rotation about the axis. The torque is negative when the force causes or tends to cause a clockwise (CW) rotation. Figure 11.3 depicts the lever arm, axis of rotation, and the direction of rotation caused by a force about some axis. The **axis of rotation** can also be called the point of torque calculation. The force can be thought of as acting anywhere along its **line of action**, shown by the dashed line. The **lever arm** (l), also called the **moment arm**, is the perpendicular distance, or shortest distance, from the line of action of the force to the axis of rotation, or point of torque calculation. The force shown in Figure 11.3 will cause or tend to cause a positive counterclockwise (CCW) rotation about the axis of rotation.

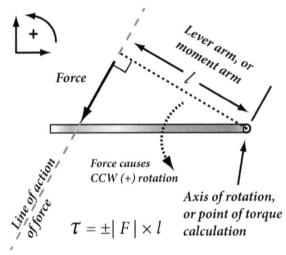

**Figure 11.3** To compute torque determine: the axis of rotation, lever arm, line of action, magnitude of force, and the direction of rotation caused by the force.

## Steps in Computing Torque

The steps in computing torque about an axis or an arbitrary point are:
1. Determine the magnitude of the force (ignore + or −).
2. Identify the line of action of the force.
3. Identify the axis, or point of torque calculation.
4. Identify the lever arm (l). The lever arm, or moment arm, is the perpendicular distance from the line of action of the force to the axis, or point of torque calculation.
5. Determine the direction of rotation (CCW or CW) that the force will cause, or tend to cause about the axis or point of torque calculation. In the formula below, the ± in front of the magnitude of the force, should be replaced with a "+" for a CCW rotation, and a "−" for a CW rotation.
6. Multiply the magnitude of the force by the lever arm to compute torque (see Figure 11.4).

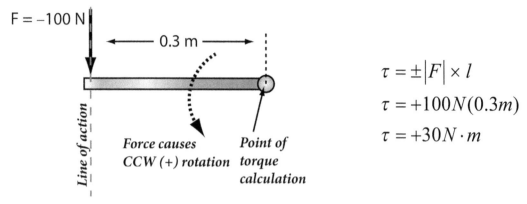

Figure 11.4 Steps in computing torque.

The −100 N force acting a perpendicular distance of 0.3 m about the point of torque calculation causes or would tend to cause +30 N·m of torque about the axis (see Figure 11.4).

## Torque Example 1

Compute the torque caused by the 100 N force about both points A and B shown in Figure 11.5.

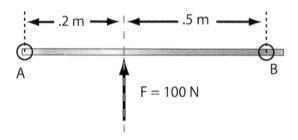

Figure 11.5 Computing the torque about points A and B.

### Solution for the Torque about Point A

The perpendicular distance (or lever arm) from the line of action of the force and point A is 0.2 m. The force will cause a CCW rotation about point A. As shown in Figure 11.6, the 100 N force cause +20 N·m of torque about point A.

### Solution for the Torque about Point B

The perpendicular distance (or lever arm) from the line of action of the force and point B is 0.5 m. The force will cause a CW rotation about point B. As shown in Figure 11.6, the 100 N force causes −50 N·m of torque about point B.

### A Force that Passes Thru the Point of Torque Calculation

A force that passes directly thru the point of torque calculation will not cause any torque about that point, since its perpendicular distance about the point is zero. In Figure 11.7,

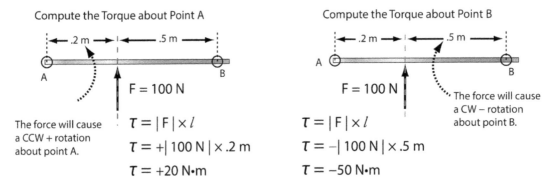

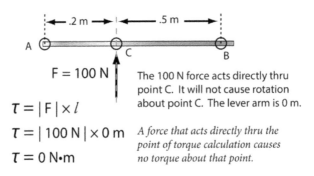

**Figure 11.6** Solutions for the torque about points A and B.

$$T = |F| \times l$$

$$T = |100\,N| \times 0\,m$$

$$T = 0\,N\cdot m$$

*A force that acts directly thru the point of torque calculation causes no torque about that point.*

**Figure 11.7** A force that acts thru the point of torque calculation (axis of rotation) causes no torque about that point. The 100 N force causes not torque about point C.

the 100 N force passes directly thru point C. Therefore, 100 N force causes 0 N·m of torque about point C.

## Torque for a Force Applied at an Angle

In Figure 11.8, the −100 N force is applied at an angle relative to the rigid object. When computing torque for a force that is applied at an angle, begin by identifying the line of action of the force. Then, after identifying the line of action of the force, determine the lever arm by identifying the perpendicular, or shortest, distance from the line of action of the force to the point of torque calculation. For example, in Figure 11.8, the lever arm about point A for the 100 N force is 0.2 m. The 100 N force causes a negative (CW) rotation about point A, which results in a torque of −20 N·m. About point B, the 100 N force has a lever arm of 0.4 m. The force causes a counterclockwise (CCW) rotation about point B. The resulting torque about point B is +40 N·m.

## Computing Torque with Two or More Forces

It is not uncommon for two or more forces to act on an object. Any force that does not pass thru the axis, or point of torque calculation, will cause torque about the axis. To compute the effects of two or more forces acting about an axis, simply compute the torque caused by each force and sum up the torques to determine the net torque. Figure 11.9 shows an example

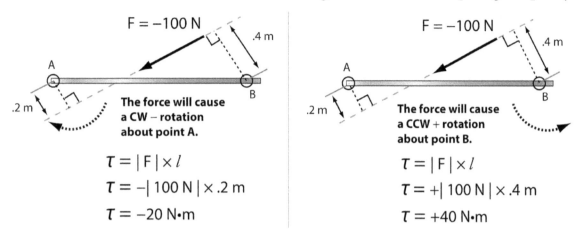

$$\tau = |F| \times l$$
$$\tau = -|100\ N| \times .2\ m$$
$$\tau = -20\ N{\cdot}m$$

$$\tau = |F| \times l$$
$$\tau = +|100\ N| \times .4\ m$$
$$\tau = +40\ N{\cdot}m$$

Figure 11.8 The −100 N force causes CW rotation about point A (left) and CCW rotation about point B (right).

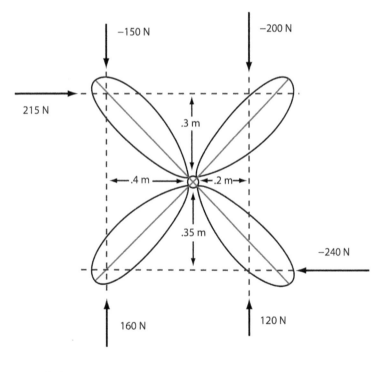

$$\sum \tau = \pm |F| \times l$$
$$\sum \tau = -160(.4) + 120(.2) - 240(.35) - 200(.2) + 150(.4) - 215(.3)$$
$$\sum \tau = -168.5 N \cdot m$$

Figure 11.9 Summing the torques caused by each force about the CM of the fan.

of computing the effects of several forces about the axis of rotation for a fan. The torque caused by each force about the axis is computed and then all of the torques are summed to determine the net torque about the axis of rotation. The forces shown in Figure 11.9 cause a net torque of $-168.5$ N·m about the axis of rotation.

## ANGULAR VERSION OF NEWTON'S LAWS OF MOTION

**I – Law of Inertia** – A body at rest stays at rest, a rotating body stays in rotation unless acted upon by a net torque.

**Inertia**, is the natural tendency of an object to remain at rest, or in motion at a constant velocity (v), in a straight line. **Angular inertia** is the natural tendency of an object to remain at rest, or in rotation at a constant angular velocity ($\omega$). **Moment of Inertia** (I) is defined as an object's resistance to angular change. Moment of inertia is a scalar. The units for moment of inertia are kg·m². An object's resistance to angular change about some axis is affected by the mass and the distance from the mass to the axis of rotation. As shown in Figure 11.10, the moment of inertia is mathematically computed by summing the product of mass times the squared distance to the axis of rotation. Figure 11.11 illustrates the effect of moving a mass closer to, or further from the axis of rotation. Both objects in Figure 11.11 have the same total mass of 25 kg. The object on the left has a greater moment of inertia and resistance to angular change, than does the object on the right. Therefore, it would be easier to spin the object on the right about the axis.

Figure 11.12 illustrates the mathematical computation of the moment of inertia about an axis of rotation, for two objects with the same mass. Each object is connected to the axis of rotation by a thin, rigid rod. The object on the left has a moment of inertia of 1.76 kg·m² about the axis of rotation and the object on the right has a moment of inertia of 2.24 kg·m² about the axis. The object on the left is easier to rotate about the origin (point o), since its moment of inertia is less than the moment of inertia for the object on the right. Conversely, the object on the right has a greater resistance to angular change than the object on the left.

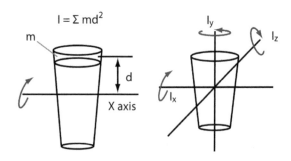

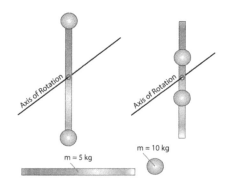

**Figure 11.10** The moment of interia (I) represents an object's resistance to angular change about some axis.

**Figure 11.11** Both objects have the same mass. The object on the left has more resistance to angular change than the object on the right.

Figure 11.13 shows an example of the moment of inertia about a somersaulting axis (medial-lateral), and a twisting axis (longitudinal) for a human. The moment of inertia about a somersaulting (medial-lateral) axis is reduced when the limbs are positioned closer to the axis of rotation, and increased when the limbs are moved further from the axis. The values shown for the moment of inertia in each position will vary based upon the actual height and mass of the individual. When changing body configuration from a layout, to a tuck position, the moment of inertia decreases from 14.0 kg·m² to 4.0 kg·m², respectively. Similarly, about a longitudinal axis, the resistance to rotation is decreased when the limbs are positioned closer to the axis and increased when positioned further from the axis.

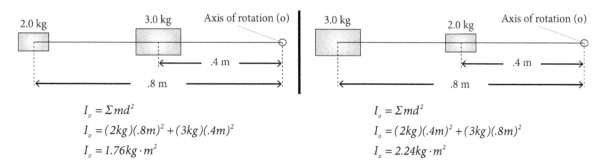

$$I_o = \Sigma m d^2$$
$$I_o = (2kg)(.8m)^2 + (3kg)(.4m)^2$$
$$I_o = 1.76 kg \cdot m^2$$

$$I_o = \Sigma m d^2$$
$$I_o = (2kg)(.4m)^2 + (3kg)(.8m)^2$$
$$I_o = 2.24 kg \cdot m^2$$

**Figure 11.12** Computation of the moment of inertia about the origin (o), for the object on the left and right.

$I_{CM} = 14.0$ kg•m²    $I_{CM} = 8.0$ kg•m²    $I_{CM} = 6.0$ kg•m²    $I_{CM} = 4.0$ kg•m²

$I_{CM} = 3.0$ kg•m²        $I_{CM} = 2.0$ kg•m²        $I_{CM} = 1.0$ kg•m²

**Figure 11.12** Examples of moment of interia about a medial-lateral (top) and longitudinal (bottom) axis.

**II – Law of Acceleration** – The angular acceleration experienced by a body is directly proportional to the net torque and inversely proportional the moment of inertia and it occurs in the direction of the net torque. The **law of acceleration** is defined by the following equation:

$$\Sigma\tau = I\alpha$$

where $\tau$ is torque in N·m, I is the moment of inertia in kg·m², and $\alpha$ is the angular acceleration in r/s². Recall from the law of inertia that, **inertia** is the natural tendency of an object to remain at rest, or in rotation at a constant angular velocity unless acted upon by a net torque. The term $I\alpha$ in the equation above represents the **angular inertia** or **inertial torques**. Inertial torques will create angular inertia ($I\omega$), which is the tendency for the object to remain in motion at a constant angular velocity ($\omega$). When computing the angular acceleration caused by a net torque, the axis of rotation must be specified. For example, to compute the angular acceleration about the center of mass (CM), use the following equation:

$$\Sigma\tau_{CM} = I_{CM}\alpha_{CM}$$

**III – Law of Reaction** – For every torque, there is an opposite and equal reaction torque.

**Law of Acceleration - Sample Problem 1**

The **law of acceleration** can be used to solve for the angular acceleration about the center of mass of a performer. Figure 11.14 shows the horizontal and vertical reaction forces for a diver during takeoff. The diver has a moment of inertia of 12 kg·m². Since neither the horizontal reaction force (Rx) or the vertical reaction force (Ry) pass thru the center of mass of the diver, they will cause torque about the center of mass of the diver. The lever arm ($l$) for Rx is given as 1.2 m, and the lever arm for Ry is 0.4 m. The horizontal reaction force (Rx) causes a CW rotation about the diver's center of mass and the vertical reaction force (Ry) causes a CCW rotation about the center of mass. **Solution:** the net torques acting about the diver's center of mass would cause an angular acceleration of 33.67 r/s² about the center of mass.

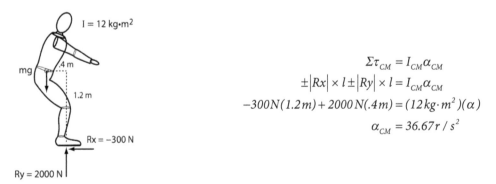

$$\Sigma\tau_{CM} = I_{CM}\alpha_{CM}$$
$$\pm|Rx| \times l \pm |Ry| \times l = I_{CM}\alpha_{CM}$$
$$-300N(1.2m) + 2000\,N(.4m) = (12\,kg\cdot m^2)(\alpha)$$
$$\alpha_{CM} = 36.67r/s^2$$

**Figure 11.14** Computation of the angular acceleration about the CM using the law of acceleration.

## Law of Acceleration - Sample Problem 2

Compute the angular acceleration (α) for the runner sprinting upstairs (see Figure 11.15). The moment of inertia about the CM is 7 kg·m². The reaction force (Rx) and the vertical reaction force (Ry) both cause torque about the center of mass, since they do not pass thru the CM. The lever arm (*l*) for Rx is given as 0.8 m, and the lever arm for Ry is 0.4 m. **Solution:** the net torques acting about the runner's CM would cause an angular acceleration of 40 r/s² about the CM.

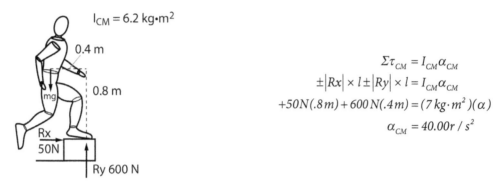

$$\Sigma\tau_{CM} = I_{CM}\alpha_{CM}$$
$$\pm|Rx| \times l \pm |Ry| \times l = I_{CM}\alpha_{CM}$$
$$+50N(.8m) + 600N(.4m) = (7\,kg\cdot m^2)(\alpha)$$
$$\alpha_{CM} = 40.00r/s^2$$

**Figure 11.15** Computation of the angular acceleration about the CM using the law of acceleration.

## Law of Acceleration - Sample Problem 3

A long jumper experiences a horizontal reaction force (Rx) of 200 N, and a vertical reaction force (Ry) of 1100 N during landing (see Figure 11.16). The moment of inertia about the CM is 6.2 kg·m². The horizontal reaction force (Rx) and the vertical reaction force (Ry) will both cause torque about the CM, since they do not pass thru the CM. The lever arm (*l*) for Rx is given as 0.4 m, and the lever arm for Ry is 0.45 m. **Solution:** the net torques acting about the runner's CM would cause an angular acceleration of −66.94 r/s² about the CM.

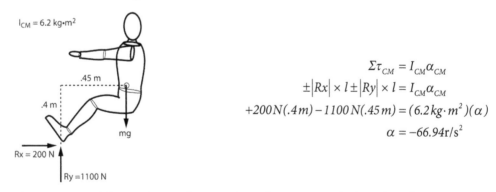

$$\Sigma\tau_{CM} = I_{CM}\alpha_{CM}$$
$$\pm|Rx| \times l \pm |Ry| \times l = I_{CM}\alpha_{CM}$$
$$+200N(.4m) - 1100N(.45m) = (6.2\,kg\cdot m^2)(\alpha)$$
$$\alpha = -66.94r/s^2$$

**Figure 11.16** Computation of the angular acceleration about the CM using the law of acceleration.

## Torque about the Elbow

Sometimes it is necessary to compute the torque about a point other than the center of mass of an object. In Figure 11.17, the bicep muscle must generate an isometric elbow flexion force to hold the 133 N weight and the forearm plus hand, which weighs 16 N. Since the force is isometric, the torque generated by the bicep must be equal and opposite to the torque caused by the dumbbell and the torque caused by the weight of the forearm and hand. In this example, the torque about the center of mass of the forearm-hand system is not relevant. Instead, it is necessary to compute the torque about the elbow joint center. Using the information shown in Figure 11.17, the bicep would be required to generate 918.4 N of force to resist the torque caused by the 133 N (30 lbs) dumbbell and the weight of the forearm and hand. When calculating the torque about the elbow, the moment of inertia must also be computed about the elbow joint rather than the center of mass of the forearm–hand segment. This is an example of a static analysis. In a static analysis, the "Iα" term on right–hand side of Newton's law of angular acceleration is zero since there is no angular acceleration.

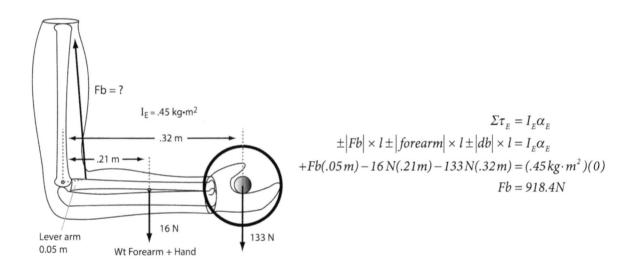

$$\Sigma \tau_E = I_E \alpha_E$$
$$\pm \left| Fb \right| \times l \pm \left| forearm \right| \times l \pm \left| db \right| \times l = I_E \alpha_E$$
$$+ Fb(.05\,m) - 16\,N(.21m) - 133\,N(.32m) = (.45\,kg \cdot m^2)(0)$$
$$Fb = 918.4N$$

Diagram labels: Fb = ?; $I_E$ = .45 kg·m²; .32 m; .21 m; Lever arm 0.05 m; 16 N; Wt Forearm + Hand; 133 N

**Figure 11.17** Example of a static torque analysis. Determine the force in the bicep (Fb) required to hold the weight.

## SUMMARY

Torque is caused by a force that does not act thru the center of mass of an object. Torque is defined as a force that causes or tends to cause rotation about a point or an axis of rotation. This chapter illustrates the steps in computing torque about some point or axis of rotation. The angular version of Newton's laws of motion are used to determine torques and angular acceleration about some axis.

## REVIEW QUESTIONS

1. Define torque and moment of inertia.
2. State Newton's laws of angular motion.
3. List the steps in computing torque.
4. Using Figure 11.18, list the lever arm for each force about points A, B, and C.
5. Using Figure 11.18, list the direction of rotation (CW or CCW) caused by each force (−50, −60, −70) about points A, B, and C.
6. Compute the sum of the torques about point A in Figure 11.18.
7. Compute the sum of the torques about point B in Figure 11.18.
8. Compute the sum of the torques about point C in Figure 11.18.
9. The diver shown in Figure 11.19 is doing a backward two-and-one-half rotation dive from a handstand position. Using Newton's Law of Angular Acceleration, complete the free-body diagram and determine the angular acceleration about the diver's center of mass.

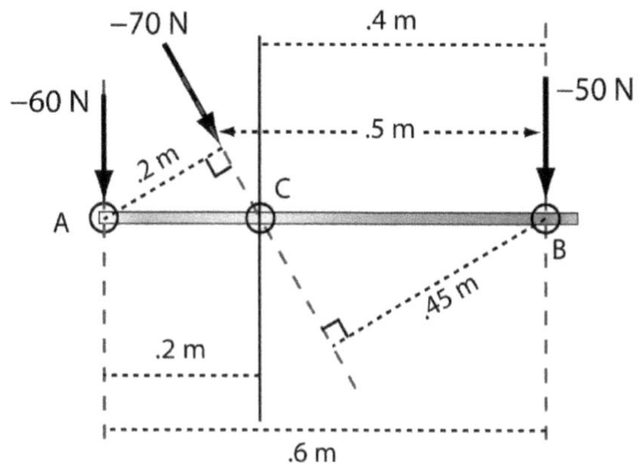

**Figure 11.18** Use this figure for questions 4-8.

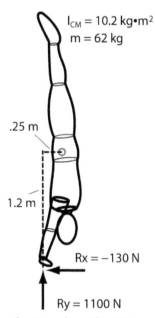

**Figure 11.19** Determine the angular acceleration about the CM.

## SUGGESTED SUPPLEMENTAL RESOURCES

Cutnell, J. D. and K. W. Johnson (2008). Physics. Hoboken, John Wiley & Sons.

Enoka, R. M. (2008). Neuromechanics of Human Movement. Champaign, Human Kinetics.

Griffiths, I. W. (2006). Principles of Biomechanics & Motion Analysis. Philadelphia, Lippincott Williams & Wilkins.

Hall, S. J. (2007). Basic Biomechanics. Boston, McGraw Hill.

Hamill, J. and K. M. Knutzen (2009). Biomechanical Basis of Human Movement. Philadelphia, Lippincott Williams & Wilkins.

Hibbeler, R. C. (2010). Engineering Mechanics: Dynamics. Upper Saddle River, Prentice Hall.

Hibbeler, R. C. (2010). Engineering Mechanics: Statics. Upper Saddle River, Prentice Hall.

McGinnis, P. M. (2013). Biomechanics of Sport and Exercise. Champaign, Human Kinetics.

Winter, D. A. (2009). Biomechanics and Motor Control of Human Movement. Hoboken, John Wiley & Sons, Inc.

# Chapter 12

## CENTER OF MASS

### STUDENT LEARNING OUTCOMES

**After reading Chapter 12, you should be able to:**
- Define the center of mass.
- Explain how to compute the center of mass of a rigid object.
- Explain how to compute the center of mass of a human.

### X COORDINATE OF THE CENTER OF MASS

The center of mass (CM) is a point that represents the average location for the total mass of a system. Figure 12.1 shows two objects of mass m1 and m2 that are located on the x axis at positions $x_1$ and $x_2$, respectively. The x location of the center of mass $X_{CM}$ is defined using the following equation.

$$x_{CM} = \frac{m_1 x_1 + m_2 x_2 + m_3 x_3 + \cdots}{m_1 + m_2 + m_3 + \cdots}$$

where the numerator of the equation is the product of each object's mass and x location from some axis or origin and the denominator is the total mass of the system. The notation "+ ⋯" indicates that the equation can be extended to account for any number of masses. If the mass of the two objects are equal then the center of mass will be located halfway between the two objects. If one of the objects is more massive the center of mass will lie closer to the more massive object.

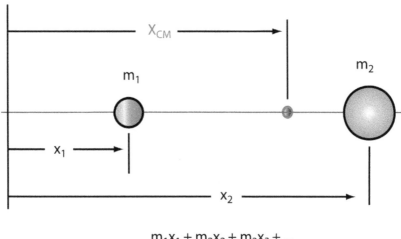

$$X_{CM} = \frac{m_1 x_1 + m_2 x_2 + m_3 x_3 + \dots}{m_1 + m_2 + m_3 + \dots}$$

**Figure 12.1** The X coordinate of the center of mass of a system of objects is located on a line between the objects, closer to the more massive object.

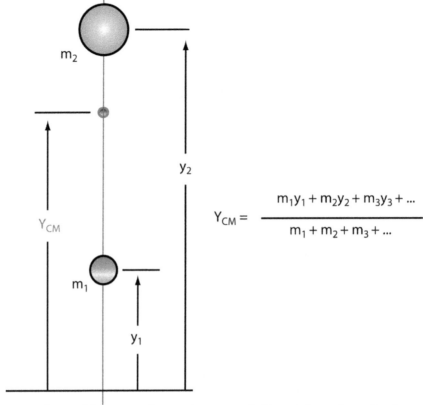

$$Y_{CM} = \frac{m_1 y_1 + m_2 y_2 + m_3 y_3 + \dots}{m_1 + m_2 + m_3 + \dots}$$

**Figure 12.2** The Y coordinate of the center of mass of a system of objects is located on a line between the objects, closer to the more massive object.

## Y COORDINATE OF THE CENTER OF MASS

The y coordinate of the center of mass of a system of objects represents the average location for the total mass of the system in the y direction. Figure 12.2 shows two objects of mass m1 and m2 that are located on the y axis at positions $y_1$ and $y_2$, respectively. The y location of the center of mass $Y_{CM}$ is defined using the following equation.

$$y_{CM} = \frac{m_1 y_1 + m_2 y_2 + m_3 y_3 + \cdots}{m_1 + m_2 + m_3 + \cdots}$$

where the numerator of the equation is the product of each object's mass and y location from some axis or origin and the denominator is the total mass of the system.

## SAMPLE CENTER OF MASS PROBLEMS

Figure 12.3 below shows a 20 kg rigid bar with a 10 kg and 25 kg block. Determine the X coordinate of the center of mass of the system of three objects.

Figure 12.4 shows the x, y locations of a system of three objects. Determine the x location of the center of mass ($X_{CM}$) and the y location of the center of mass ($Y_{CM}$) of the system.

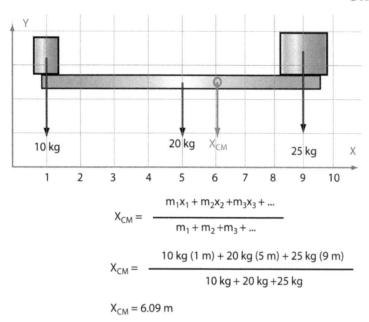

$$X_{CM} = \frac{m_1x_1 + m_2x_2 + m_3x_3 + \ldots}{m_1 + m_2 + m_3 + \ldots}$$

$$X_{CM} = \frac{10 \text{ kg } (1 \text{ m}) + 20 \text{ kg } (5 \text{ m}) + 25 \text{ kg } (9 \text{ m})}{10 \text{ kg} + 20 \text{ kg} + 25 \text{ kg}}$$

$$X_{CM} = 6.09 \text{ m}$$

**Figure 12.3** Compute the X coordinate of the center of mass of the system of objects.

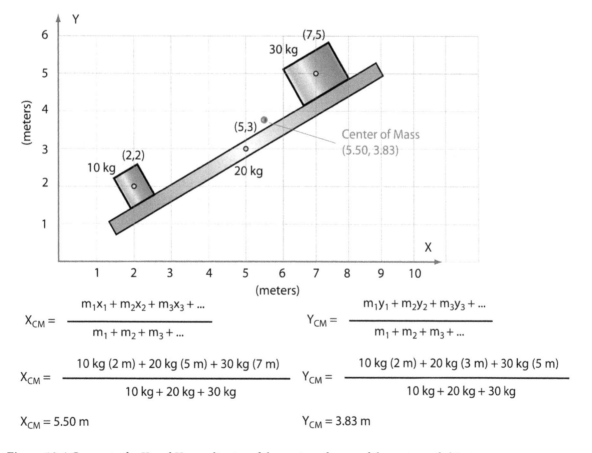

$$X_{CM} = \frac{m_1x_1 + m_2x_2 + m_3x_3 + \ldots}{m_1 + m_2 + m_3 + \ldots} \qquad Y_{CM} = \frac{m_1y_1 + m_2y_2 + m_3y_3 + \ldots}{m_1 + m_2 + m_3 + \ldots}$$

$$X_{CM} = \frac{10 \text{ kg } (2 \text{ m}) + 20 \text{ kg } (5 \text{ m}) + 30 \text{ kg } (7 \text{ m})}{10 \text{ kg} + 20 \text{ kg} + 30 \text{ kg}} \qquad Y_{CM} = \frac{10 \text{ kg } (2 \text{ m}) + 20 \text{ kg } (3 \text{ m}) + 30 \text{ kg } (5 \text{ m})}{10 \text{ kg} + 20 \text{ kg} + 30 \text{ kg}}$$

$$X_{CM} = 5.50 \text{ m} \qquad Y_{CM} = 3.83 \text{ m}$$

**Figure 12.4** Compute the X and Y coordinates of the center of mass of the system of objects.

## SEGMENTAL METHOD FOR COMPUTING THE CENTER OF MASS

An engineer building a machine with two or more rigid links can easily determine the mass, location of the center of mass and the moment of inertia of each rigid link. The engineer can simply dismantle the machine to weigh each component, balance the segment on a knife edge to determine the location of the center of mass, and spin the link about a thin rod thru the center of mass to determine the moment of inertia. Then after determining the location of the center of mass, and moment of inertia the engineer can rebuild the machine.

In biomechanics our machine is the human. For obvious reasons we are not able to dismantle the human to determine the: mass of each body part, location of each segments center of mass, and moment of inertia about all three anatomical axes. As a result we have relied upon detailed measurements of cadavers to develop prediction equations to estimate the mass of a body segment, the location of the center of mass and moment of inertia about all three anatomical axes. The earliest recorded work in human anthropometrics was done in the 17th century by Borelli (Borelli, 1680). He determined the center of mass of the body by having men lie on a rigid platform that was supported by a knife edge. In 1889, Braune and Fisher determined the weight, volume and center of mass of human body segments by dissecting three adult male cadavers (Braune, et al., 1889). Several detailed anthropometric studies were conducted for the Air Force and NASA at Wright-Patterson Air Force Base (Chandler, et al., 1975; Clauser, et al., 1969; Dempster, 1955; Hanavan, 1964). In addition to using cadaver dissection to obtain body segment parameter some researchers have used mathematical modeling along with anthropometric measurements (Dainis, 1980; Hanavan, 1964; Hatze, 1980; Vaughan, et al., 1999). Finally, Zatsiorsky and co-workers determined body segment parameters of living subjects using gamma mass scanning (Zatsiorsky, et al., 1983; Zatsiorsky, et al., 1985; Zatsiorsky, et al., 1991; Zatsiorsky, et al., 1990).

The segmental method of computing the center of mass for a human involves computing the center of mass of individual body segments and then the location of the center of mass of the body is found by summing the torques produced by each body segment about the X and Y axes and dividing the sum of the torques about each axis by the total mass of the system. The steps necessary to find the center of mass of the body are:

1.  Determine the X, Y location of the end points of all body segments.
2.  Determine the mass of each segment.
3.  Determine the X, Y location of each segments center of mass.

Once the X, Y locations of each segments center of mass have be determined the total body center of mass can be computed using the center of mass equations:

$$x_{CM} = \frac{m_1 x_1 + m_2 x_2 + m_3 x_3 + \cdots}{m_1 + m_2 + m_3 + \cdots} \qquad y_{CM} = \frac{m_1 y_1 + m_2 y_2 + m_3 y_3 + \cdots}{m_1 + m_2 + m_3 + \cdots}$$

Table 12.1 gives prediction equations that can be used to estimate a body segments weight from body weight and the location of each segments center of mass from its proximal end. This data set is based upon the work of Chandler and co-workers (Chandler, et al., 1975). Figure 12.5 gives an example of using the equations in Table 12.1 to determine an individual segments center of mass location for the forearm segment.

**Table 12.1** Segment Weight Prediction Equations and Center of Mass Locations

| Segment | Segment Weight (N) | CM Location (%) | Proximal End |
|---|---|---|---|
| Head | $0.032 \times BW + 18.7$ | 66.3 | Top of Head |
| Trunk | $0.532 \times BW - 6.93$ | 52.2 | 1st Cervical |
| Upper Arm | $0.022 \times BW + 4.76$ | 50.7 | Shoulder |
| Forearm | $0.013 \times BW + 2.41$ | 41.7 | Elbow |
| Hand | $0.005 \times BW + 0.75$ | 51.1 | Wrist |
| Thigh | $0.127 \times BW - 14.82$ | 39.8 | Hip |
| Lower Leg | $0.044 \times BW - 1.75$ | 41.3 | Knee |
| Foot | $0.009 \times BW + 2.48$ | 40.0 | Ankle |

From: Chandler, R.F., C.E. Clauser, et al. (1975) Investigation of Inertial Properties of the Human Body (AM-RL-TR-74-137). Wright-Patterson Air Force Base. Dayton, OH, Aerospace Medical Division NTIS No. AD-A016 485.

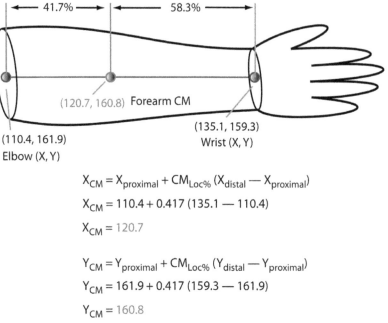

$$X_{CM} = X_{proximal} + CM_{Loc\%} (X_{distal} - X_{proximal})$$
$$X_{CM} = 110.4 + 0.417 (135.1 - 110.4)$$
$$X_{CM} = 120.7$$

$$Y_{CM} = Y_{proximal} + CM_{Loc\%} (Y_{distal} - Y_{proximal})$$
$$Y_{CM} = 161.9 + 0.417 (159.3 - 161.9)$$
$$Y_{CM} = 160.8$$

Figure 12.5 Example using the equations in Table 12.1 to determine the X, Y location of the center of mass for the forearm segment.

The example below in Figure 12.6 shows how to compute the center of mass of the body using the segmental method, when given segment weights and the individual segment center of mass X, Y locations. The subject in Figure 12.6 has a weight of 756 N and a height of 185 cm. The right and left sides of the upper and lower limbs are symmetrical in this figure, and as a result, the weights for the upper and lower limbs have been doubled.

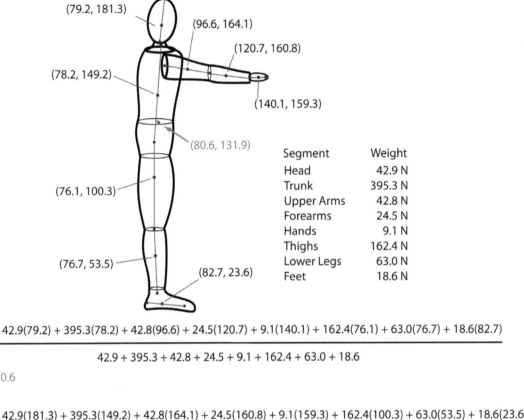

| Segment | Weight |
|---|---|
| Head | 42.9 N |
| Trunk | 395.3 N |
| Upper Arms | 42.8 N |
| Forearms | 24.5 N |
| Hands | 9.1 N |
| Thighs | 162.4 N |
| Lower Legs | 63.0 N |
| Feet | 18.6 N |

$$X_{CM} = \frac{42.9(79.2) + 395.3(78.2) + 42.8(96.6) + 24.5(120.7) + 9.1(140.1) + 162.4(76.1) + 63.0(76.7) + 18.6(82.7)}{42.9 + 395.3 + 42.8 + 24.5 + 9.1 + 162.4 + 63.0 + 18.6}$$

$$X_{CM} = 80.6$$

$$Y_{CM} = \frac{42.9(181.3) + 395.3(149.2) + 42.8(164.1) + 24.5(160.8) + 9.1(159.3) + 162.4(100.3) + 63.0(53.5) + 18.6(23.6)}{42.9 + 395.3 + 42.8 + 24.5 + 9.1 + 162.4 + 63.0 + 18.6}$$

$$Y_{CM} = 131.9$$

**Figure 12.6** Example of using the segmental method to determine the X, Y location of the body center of mass.

## REFERENCES

Borelli GA. De motu animalium: Lugduni Batavorum; 1680.

Braune W, Fischer O. The center of gravity of the human body as related to the german infantryman. In: Center DD, editor. Leipzig, 1889.

Chandler RF, Clauser CE, McConville JT, Reynolds HM. Investigation of inertial properties of the human body (amrl-tr-74-137). In: Base W-PAF, editor. Dayton, OH: Aerospace Medical Division NTIS No. AD-A016 485, 1975.

Clauser CE, McConville JT, Young JW. Weight, volume, and center of mass of segments of the human body (amrl-tr-69-70). In: Laboratory AMR, editor. Dayton, OH: Wright-Patterson Air Force Base. p. 101, 1969.

Dainis A. Whole body and segment center of mass determination from kinematic data. Journal of Biomechanics. 13: 647-651, 1980.

Dempster WT. Space requirements of the seated operator. In: Laboratory AMR, editor. Dayton, OH: Wright-Patterson Air Force Base (WADC TR 55-199). p. 254, 1955.

Hanavan EP. A mathematical model of the human body. In: Laboratory AMR, editor. Dayton, OH: Wright-Patterson Air Force Base, 1964.

Hatze H. A mathematical model for the computational determination of parameter values of anthropomorphic segments. Journal of Biomechanics. 13: 833-843, 1980.

Vaughan CL, Davis BL, O'Connor JC. Dynamics of human gait. Cape Town: Kiboho; 1999.

Zatsiorsky V, Seluyanov V. The mass and inertia characteristics of the main segments of the body. In: Matsui H, Kobayashi K, editors. Biomechanics viii-b. Champaign: Human Kinetics. p. 1152-1159, 1983.

Zatsiorsky V, Seluyanov V. Estimation of the mass and inertia characteristics of the human body by means of the best predictive regression equation. In: Winter DA, Norman RW, Wells RP, Hayes KC, Patla AE, editors. Biomechanics ix-b. Champaign: Human Kinetics. p. 233-239, 1985.

Zatsiorsky VM, Chugunova LG. Methods of determining mass-inertial characteristics of human body segments. In: Chernyi GG, Regirer SA, editors. Contemporary problems of biomechanics. Moscow: Mir Publishers. p. 272-291, 1991.

Zatsiorsky VM, Seluyanov V, Chugunova L. In vivo body segment inertial parameters determination using a gamma-scanner method. In: Berme N, Cappozzo A, editors. Biomechanics of human movement: Applications in rehabilitation, sports and ergonomics. Worthington, OH: Bertec Corporation. p. 186-202, 1990.

## SUGGESTED SUPPLEMENTAL SOURCES

Cutnell, J. D. and K. W. Johnson (2008). Physics. Hoboken, John Wiley & Sons.

Enoka, R. M. (2008). Neuromechanics of Human Movement. Champaign, Human Kinetics.

Griffiths, I. W. (2006). Principles of Biomechanics & Motion Analysis. Philadelphia, Lippincott Williams & Wilkins.

Hall, S. J. (2007). Basic Biomechanics. Boston, McGraw Hill.

Hamill, J. and K. M. Knutzen (2009). Biomechanical Basis of Human Movement. Philadelphia, Lippincott Williams & Wilkins.

McGinnis, P. M. (2013). Biomechanics of Sport and Exercise. Champaign, Human Kinetics.

# Chapter 13

## ANGULAR IMPULSE

### STUDENT LEARNING OUTCOMES

**After reading Chapter 13, you should be able to:**
- Define angular impulse and angular momentum.
- Define the angular impulse–angular momentum relationship.
- Define the **principle of conservation of angular momentum**.

### ANGULAR MOMENTUM

In chapter 8, linear momentum (p) was defined as the object's mass (m) times its velocity (v) using the following equation:

$$p = mv$$

**Angular momentum** (L) is the quantity of rotation a body possesses. Angular momentum can be defined as the product of the object's moment of inertia ($I$) times its angular velocity ($\omega$). Angular momentum is calculated as follows:

$$L = I\omega$$

Where (L) is the angular momentum has units of (kg·m²/s), ($I$) is the moment of inertia has units of kg·m², and ($\omega$) is the angular velocity in r/s. Angular momentum is a vector that is defined by the right-hand rule. The direction of the angular momentum vector is the same as the direction of the angular velocity vector (CCW is positive, CW is negative). Figure 13.1 shows an example of how angular momentum is computed. The gymnast in Figure 13.1 has an angular velocity of –4.2 r/s and a moment of inertia of 5.0 kg·m². The resulting angular momentum is negative (clockwise) and has a magnitude of –21 kg·m²/s.

Angular momentum is generated or absorbed by the application of torques. Recall from the angular version of Newton's **law of inertia** that an object at rest stays at rest and an object that is rotating will continue to rotate unless acted upon by a net torque. Therefore, to cause an object to rotate (create angular momentum) or to stop an object from rotating (absorb angular momentum), you must apply a torque. The application of a torque will change an object's angular inertia ($I\alpha$). Newton's **law of angular acceleration** can be used to quantify the effect of a net torque on creating or arresting angular inertia. According to the **law of angular acceleration**, the angular acceleration ($\alpha$) that an object experiences is directly

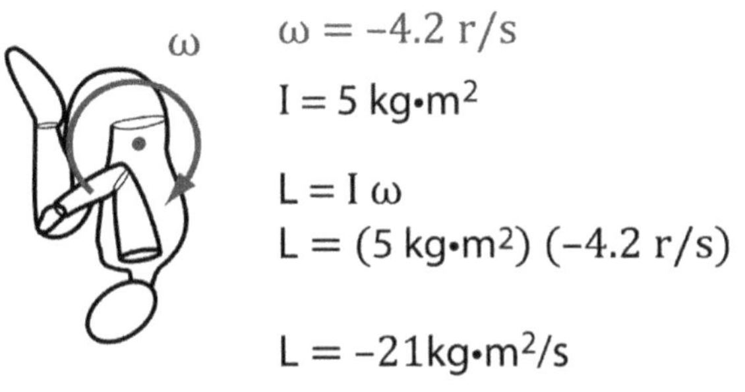

$$\omega = -4.2 \text{ r/s}$$
$$I = 5 \text{ kg·m}^2$$
$$L = I\omega$$
$$L = (5 \text{ kg·m}^2)(-4.2 \text{ r/s})$$
$$L = -21 \text{kg·m}^2\text{/s}$$

**Figure 13.1** Angular momentum represents the quantity of rotation a body possesses.

149

proportional to the net torque ($\tau$) and inversely proportional to the object's moment of inertia ($I$).

$$\Sigma\tau = I\alpha$$

## ANGULAR IMPULSE-ANGULAR MOMENTUM RELATIONSHIP

Recall from chapter 8 that impulse (J) was defined as torque times time, or the area under the torque-time graph. Angular impulse ($\Sigma\tau(t)$) is the angular version of linear impulse and it is simply the torque times time, or the area under the torque-time graph. Angular impulse has units of N·m·s and it is a vector defined by the right-hand rule. The relationship between angular impulse and angular momentum can be derived by substituting the angular acceleration equation into Newton's **law of angular acceleration** as follows:

$$\alpha = \frac{\omega_f - \omega_i}{\Delta t}$$

$$\Sigma\tau = I\alpha$$

$$\Sigma\tau = I\left(\frac{\omega_f - \omega_i}{t}\right)$$

$$\Sigma\tau(t) = I\omega_f - I\omega_i$$

In words, the last equation indicates that the angular impulse (torque × time) is equal to the change in angular momentum ($I\omega_f - I\omega_i$). According to the angular impulse–angular momentum equation, when you apply a net torque over some time interval, you will change the angular momentum of an object.

### Example Angular Impulse Problem

The rigid bar shown in Figure 13.2 has a moment of inertia ($I$) of 0.8 kg·m² and an initial angular velocity ($\omega_i$) of 0 r/s. A constant torque of 4.55 N·m is applied for $t = 0.92$ s. Compute the final angular velocity ($\omega_f$) and the angular impulse that was applied to the rigid bar. As shown in Figure 13.2, the angular impulse was 4.186 N·m·s. The angular impulse causes the angular velocity to increase from 0 r/s to 5.23 r/s. The resulting angular momentum generated by this angular impulse can be computed as follows:

$$L = I\omega$$

$$L = (0.8 kg \cdot m^2)(5.23 r/s)$$

$$L = 4.18 kg \cdot m^2/s$$

$\tau = 4.55$ N·m
$I = 0.8$ kg·m²
$t = 0.92$ s
$\omega_i = 0.0$ r/s

$$\Sigma\tau(t) = I\omega_f - I\omega_i$$
$$4.55(0.92) = 0.8(\omega_f) - 0.8(0)$$
$$4.186 \text{ N·m·s} = 0.8 \text{ kg·m}^2 (\omega_f)$$
$$\omega_f = \frac{4.186 \text{ N·m·s}}{0.8 \text{ kg·m}^2}$$
$$\omega_f = 5.23 \text{ r/s}$$

**Figure 13.2** An angular impulse is applied to the rigid bar, changing its angular momentum.

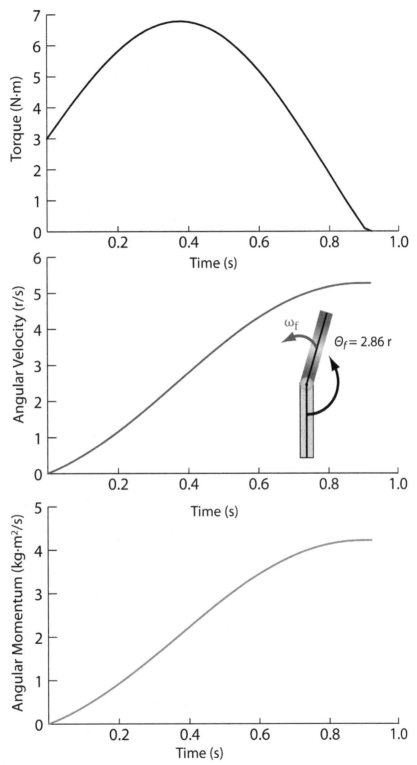

**Figure 13.3** Angular impulse for a non-constant torque is computed by calculating the area underneath the torque-time curve. The angular impulse, which is the area under the torque-time curve (top graph), is equal to the change in angular momentum (bottom graph).

The example shown in Figure 13.2 computes the angular impulse for a constant torque that is applied over some time interval. To compute the angular impulse for a non-constant torque, the torque must be integrated with respect to time. Figure 13.3 gives an example of a non-constant torque. The angular impulse for a non-constant torque is computed by integrating the torque as follows:

$$\int_{t0}^{t1} \tau\ dt = I\omega_f - I\omega_i$$

The **angular impulse** for a non-constant torque is defined as the area underneath the torque-time curve. The units for angular impulse are N·m·s. According to the **angular impulse-angular momentum relationship**, the angular impulse is equal to the change in angular momentum. In Figure 13.3, the torque is applied to a rigid bar, which has a moment of inertia ($I$) of 0.8 kg·m$^2$, about its axis of rotation. The angular velocity curve is shown in the middle graph of Figure 13.3. Prior to the application of the torque, the angular velocity ($\omega$) of the bar was 0 r/s. After the torque was applied, the angular velocity increased to 5.27 r/s. The angular momentum graph, shown in the bottom of Figure 13.3, has the same shape as the angular velocity graph above it. A rigid object with only one segment is unable to change its configuration, and therefore, the moment of inertia of the rigid bar is constant at 0.8 kg·m$^2$ throughout the movement. Unlike the rigid bar, humans are able to change body configuration. A change in body configuration will cause an increase in the moment of inertia, if the body segments are moved away from the axis of rotation, or it will cause a decrease in moment of inertia, if the body segments are moved closer to the axis of rotation. Figures 11.10-11.12, in chapter 11, illustrate the effect of alterations of body segment position relative to the axis of rotation on moment of inertia. The next section will describe the **principle of conservation of angular momentum** and the relationship between angular velocity and moment of inertia on angular momentum in the air.

## Principle of Conservation of Angular Momentum

Angular momentum can be increased or decreased by applying torques over time. When an object or performer is in the air, it is not possible to apply a torque, since there is nothing to push or pull against, and as a result, angular momentum is constant in the air. This is known as the **principle of conservation of angular momentum**. According to the **principle of conservation of angular momentum**, angular momentum is constant in the air because gravity is the only force acting on the system and the net external torques are zero. Additionally, it can be stated that angular momentum is constant in the air because angular impulse is zero in the air.

The relationship between angular momentum (L), moment of inertia ($I$), and angular velocity ($\omega$) for a diver in the air is shown in Figure 13.4. The angular momentum in Figure 13.4 is constant at 50 kg·m$^2$/s throughout the flight of the dive. When the diver goes from a layout position at takeoff, to a tuck at the peak, this reduces the moment of inertia from 15 kg·m$^2$ to 3 kg·m$^2$. This reduction of moment of inertia causes the diver to spin faster, and as a result, angular velocity ($\omega$) increases from 3.33 r/s at takeoff, to 16.67 r/s when the diver is in the tuck position. Shortly after the peak, the diver opens up from the tuck position to a layout

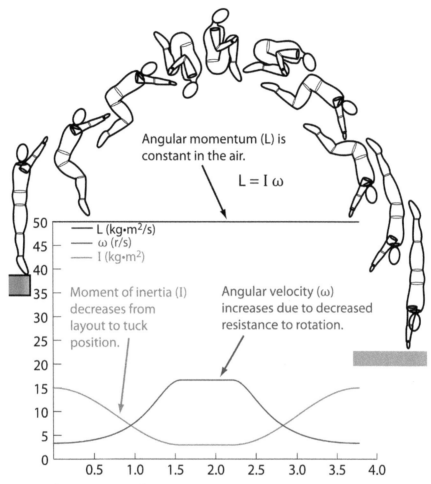

Angular momentum (L) is constant in the air.

$$L = I \, \omega$$

Moment of inertia (I) decreases from layout to tuck position.

Angular velocity (ω) increases due to decreased resistance to rotation.

L (kg·m²/s)
ω (r/s)
I (kg·m²)

**Figure 13.4 Principle of conservation of angular momentum**. In the air, when gravity is the only force operating, angular momentum is constant.

position, which causes the moment of inertia to increase and the angular velocity to decrease. At any time point in the flight, if you multiply the angular velocity by the moment of inertia, you will get the constant value of 50 kg·m²/s for the angular momentum. Once the diver is airborne, he or she, cannot change the magnitude of angular momentum during the flight phase. Therefore, the diver must attain sufficient angular momentum to complete the rotation prior to takeoff. If the diver does not have enough angular momentum to complete the dive, he or she can compensate by tucking earlier, or holding the tuck longer to increase the angular velocity and angular inertia ($I\omega$). Similarly, if the diver has too much rotation, he or she can tuck less, or delay going into the tuck, to reduce angular velocity, and slow down rotational inertia ($I\omega$)

The **principle of conservation of angular momentum** can also be observed by watching a skater execute a triple axel in skating. Just prior to takeoff, the skater has his or her arms stretched out. After takeoff, the skater pulls their arms closer to the body, which reduces the moment of inertia about the longitudinal axis and causes the skater's angular velocity to increase. Then, as the skater prepares to land, he or she will open their arms up, which causes

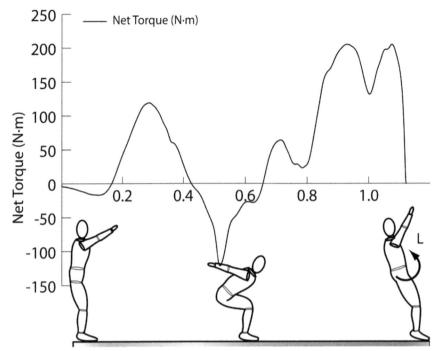

**Figure 13.5** The application of a net torque over some time interval creates backward (CCW) angular momentum (L) for a backward somersault.

the moment of inertia to increase, and the angular velocity to decrease. Again, at any point in the flight the skater's angular momentum about the longitudinal axis is constant.

Whenever a gymnast or diver performs a somersaulting motion, they must generate the necessary angular momentum to complete the rotation prior to leaving the ground. Figure 13.5 shows a gymnast pushing against the ground to execute a back flip. Initially, the gymnast's angular velocity is zero. While in contact with ground, the gymnast applies forces against the ground. If the reaction forces do not act through the center of mass of the gymnast, they will cause torque about the CM. In Figure 13.5, the net torque is positive from 0.66 s until takeoff, at 1.12 s. This positive torque in Figure 13.5 will generate positive (CCW) angular momentum (L), and positive angular velocity ($\omega$) at takeoff (see Figure 13.6). Since the gymnast is in contact with the ground, he or she can change angular momentum by applying torque over time (angular impulse). Notice that the angular momentum in Figure 13.6 is not constant. The change in angular momentum in Figure 13.6, is directly related to the angular impulse that the gymnast applied during ground contact. Refer back to Figure 13.5. In Figure 13.5, the area of positive impulse is greater than the area of negative impulse. This net positive angular impulse enables the gymnast to attain 62 kg·m²/s of angular momentum at takeoff.

Once the gymnast is in the air, the angular momentum is constant (see Figure 13.7). In the Figure 13.7, the angular velocity ($\omega$) increases, and the moment of interia ($I$) decreases from takeoff to the peak, as the gymnast goes from the layout position at takeoff, to the tuck position at the peak. Then from the peak to the landing, the gymnast opens from the tuck to the layout position, and this causes the moment of inertia to increase and the angular velocity to decrease. The torques generated in Figure 13.5 created the positive angular momentum at

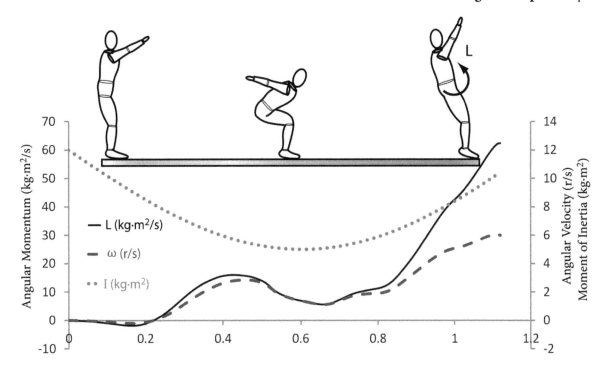

**Figure 13.6** Change in angular momentum (L) during ground contact prior to executing a backward somersalt.

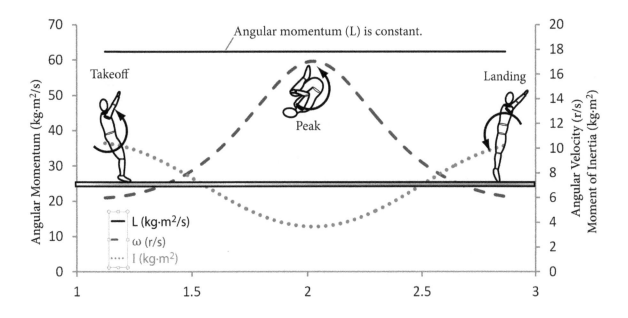

**Figure 13.7** During the flight phase of the backward somersault, the gymnast's angular momentum (L) is constant. From takeoff to peak, the gymnast changes body configuration from a layout to a tuck, which reduces moment of inertia and increases angular velocity. This pattern is reversed from peak to landing, causing an increase in moment of inertia and a decrease in angular velocity.

takeoff, shown in Figure 13.6. Then in Figure 13.7, the gymnast is in the air, and according to the **principle conservation of angular momentum,** the angular momentum is constant throughout the flight phase.

## COMPUTING ANGULAR MOMENTUM WITH VARYING MOMENT OF INERTIA

This example will use the Excel file "**Back Flip Practice File.xls**", which is available on the course website. In this example, we will compute the angular velocity, and angular momentum for the back flip motion graphed above in Figures 13.5-13.7. In this motion, the gymnast is changing body position so the moment of inertia is not constant. First, use the angular impulse–angular momentum relationship to compute the change in angular velocity.

$$\int_{t0}^{t1} \tau \, dt = I\omega_f - I\omega_i$$

Rearrange the above equation to solve for final angular velocity.

$$\omega_f = \frac{\int_{t0}^{t1} \tau \, dt + I\omega_i}{I}$$

Now convert to the above equation into Excel format and place it in cell C3. Copy the formula down the rest of column C to compute a continuous angular velocity curve. Note, the time between data points in this file is 1/1080 seconds.

=((B3*(1/1080))+(C2*D2))/D3

Now compute angular momentum as follows:
$$L = I\omega$$
Place the following equation in cell E2 to translate the above equation into an Excel equation.

=C2*D2

The angular impulse is the area under the torque–time curve. To get angular impulse, first multiply the torque x time using the following equation:

=B2*(1/1080)

Place the above equation in cell F2, and then copy it down the column. Now get the angular impulse from t = .657 (row 712) until the end of the curve at t = 1.12 (row 1212) by summing column F, from rows 712 to 1212 using the following formula:

=SUM(F712:F1212)

Place the above formula in cell J2. The angular impulse over this time interval is 56.73 N·m·s. This positive angular impulse enables the gymnast to obtain 62.39 kg·m$^2$/s of angular momen-

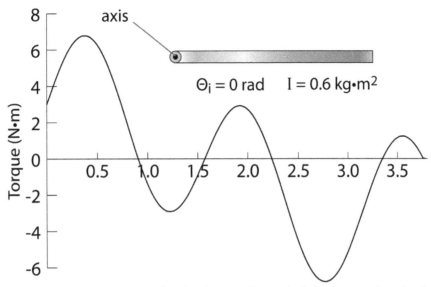

**Figure 13.8** A non-constant torque is applied to the thin rigid bar, which has an initial angle of 0.0 rad, and a moment of inertia of 0.6 kg·m². Compute the resulting angular momentum and angular velocity in Excel.

tum at takeoff. The angular momentum is positive, indicating that the gymnast will rotate in the CCW direction.

## COMPUTING ANGULAR MOMENTUM WITH CONSTANT MOMENT OF INERTIA

The thin rigid bar shown in Figure 13.8 has a constant moment of inertia of 0.6 kg·m² about the axis of rotation. The initial angular position of the bar to the right horizontal is 0 radians, and the initial angular velocity is 0 r/s. A torque motor is used to apply the torque to the bar shown in Figure 13.8. Compute the following variables as a function of time: angular velocity, angular momentum, and angle. Compute the angular impulse for all positive and negative phases of the torque–time graph. Use the Excel file "**Rigid Bar.xls**", which is available on the course website.

Place the following formula in cell C3, and copy down the column to compute the angular velocity using the angular impulse–angular momentum relationship (Note, the moment of inertia is located in cell H2, so the Excel fixed reference method "$H$2" is used to get the moment of inertia):

$$=((B3*0.02)+(C2*\$H\$2))/\$H\$2$$

Now compute the angular momentum by placing the following formula in cell D2 and copying down the column D:

$$=C2*\$H\$2$$

Now compute the torque × time so that it can be summed later to compute the angular impulse for all positive and negative phases of the torque–time curve. Place the following equation in cell E2, and copy it down column E (Note the time between data points is 0.02 seconds):

$$=B2*0.02$$

Recall from chapter 6 that angular velocity is the rate of change in angular position.

$$\omega = \frac{\theta f - \theta i}{\Delta t}$$

The above equation can be rearranged to compute the final angle as a function of time, from the initial angle, and the angular velocity.

$$\theta_f = \omega t + \theta_i$$

The equation above can be translated into the following Excel-friendly equation, which should be placed in cell F3 and copied down column F:

$$=(C3*0.02)+F2$$

Now compute the angular impulse for each of the positive and negative phases of the torque–time curve by summing the appropriate rows in column E. For example, to compute the angular impulse of the first positive phase of the curve, sum rows E2 to E47 using the following equation:

$$=SUM(E2:E47)$$

Figure 13.9 shows the computed angular impulses for each phase.

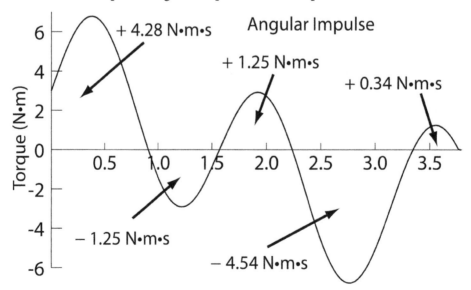

**Figure 13.9** Angular impulses produced by each positive and negative torque in Figure 13.8.

## ANGULAR MOMENTUM IN THE X GAMES

If you happen to be a fan of the X Games, you may of already seen this: back in 2006, a motocross rider by the name of Travis Pastrana did the first double back flip on a motorcycle. So now that you know a little about angular momentum, go to Google or YouTube and search for "Travis Pastrana double back flip". After viewing a double back flip on a motorcycle, consider the limitations of doing a flip on a motorcycle. The bike is very heavy, so its moment of inertia (resistance to rotation) is high. Unlike diving or gymnastics, you can't fold (tuck, pike) the bike to reduce its moment of inertia. Therefore, the jumper must generate sufficient angular momentum prior to leaving the ramp. Once in the air, there is very little that the jumper can do to increase, or decrease angular velocity. To reduce angular velocity the jumper can extend the body as far away from the bike as possible. To increase angular velocity, the jumper can move the body as close as possible to the center of mass of the bike. However, both of these adjustments will have a minimal effect on changing the rotational inertia ($I\omega$), since the mass of the jumper is relatively small in comparison to the mass of the jumper-plus-bike system.

## REVIEW QUESTIONS

1. A diver in air has 70 kg·m²/s of angular momentum and a moment of inertia of 12 kg·m². What is his angular velocity? If the diver reduces his moment of inertia to 6 kg·m² while in the air how does this affect angular momentum, and angular velocity?
2. An X games, athlete applies a torque of 30 N·m for a time of t = 0.2 seconds while in a layout position with a moment of inertia $I$ = 10 kg·m². His initial angular velocity was $\omega_i$ = 0.3 r/s. What is his angular velocity after experiencing this torque? What angular impulse was applied?
3. A diver with an initial angular velocity of –0.5 r/s and a moment of inertia of 7 kg·m² applies forces against the board, which produces an angular impulse of 21 N·m·s. What is his final angular velocity after applying this impulse?
4. Write the angular impulse–angular momentum equation. State the equation in words. What are the units for angular impulse and angular momentum? What is angular impulse? What is angular momentum? What is moment of inertia, and what are the units of moment of inertia?
5. What is the **principle of conservation of angular momentum**? How do you create angular momentum? When can you change angular momentum? Draw a picture to show the relationship between angular momentum, angular velocity, and moment of inertia for a diver in the air who goes from a layout to a tuck and back to a layout before entering the water.
6. A torque of –15 N·m is applied for t = 0.7 seconds on a rigid bar with an initial angular velocity of $\omega_i$ = +0.2 r/s and a moment of inertia of 10 kg·m². What was the final angular velocity after this torque was applied? What angular impulse was applied?
7. Download the Excel file "**Back Flip Practice File.xls**" and compute the following: angular velocity, angular momentum, angular impulse for all positive and negative areas of the torque–time curve, angular position.

8. Download the Excel file: "**Rigid Bar.xls**" and compute the following: angular velocity, angular momentum, angular impulse for all positive and negative areas of the torque–time curve, angular position.

## SUGGESTED SUPPLEMENTAL SOURCES

Cutnell, J. D. and K. W. Johnson (2008). Physics. Hoboken, John Wiley & Sons.

Enoka, R. M. (2008). Neuromechanics of Human Movement. Champaign, Human Kinetics.

Griffiths, I. W. (2006). Principles of Biomechanics & Motion Analysis. Philadelphia, Lippincott Williams & Wilkins.

Hall, S. J. (2007). Basic Biomechanics. Boston, McGraw Hill.

Hamill, J. and K. M. Knutzen (2009). Biomechanical Basis of Human Movement. Philadelphia, Lippincott Williams & Wilkins.

Hibbeler, R. C. (2010). Engineering Mechanics: Dynamics. Upper Saddle River, Prentice Hall.

Hibbeler, R. C. (2010). Engineering Mechanics: Statics. Upper Saddle River, Prentice Hall.

McGinnis, P. M. (2013). Biomechanics of Sport and Exercise. Champaign, Human Kinetics.

Winter, D. A. (2009). Biomechanics and Motor Control of Human Movement. Hoboken, John Wiley & Sons, Inc.

# Chapter 14

## MECHANICAL PROPERTIES OF MATERIALS

### LEARNING OUTCOMES

**After reading Chapter 14, you should be able to**

- Define stress, strain, stiffness, and compliance.
- Describe the stress-strain curves for ductile, brittle, and elastic materials.
- Define viscoelasticity, viscosity, damping, plasticity, and elasticity.
- Describe the difference between an elastic response and a viscoelastic response.
- Define Young's modulus.
- Define hysteresis, energy absorbed, energy returned, creep, and force relaxation.
- Describe the effects of temperature and velocity of loading on the mechanical response of viscoelastic materials.

### STRENGTH OF MATERIALS AND TYPES OF LOADS

The strength of a biological material can be defined by the ability of the material to withstand forces without breaking or failing. Several factors such as: microstructure, age, temperature, fluid content, type, velocity, and direction of loading can alter the strength of a material. Figure 14.1 shows the types of loads that can be applied to a material. Compression and tension are both axial loads, since the forces are directed along the long axis of the material. Torsion and bending are known as shear loads, since the forces act in a tangential or parallel direction.

When a load is applied to a material, the material will develop internal resistance to the applied load. The magnitude of this internal resistance to applied loads is dependent upon the

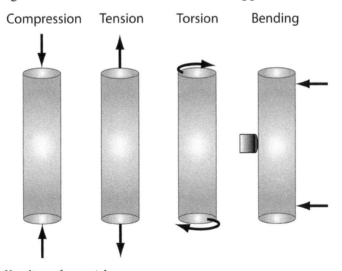

**Figure 14.1** Direction of loading of materials.

stiffness of the material. For example, the internal resistance developed by compressing a tennis ball is considerably less than the internal resistance developed by compressing a bowling ball.

## Axial Stress

The internal resistance of a material to an applied axial load is called axial stress ($\sigma$). The Greek symbol sigma ($\sigma$) is typically used for axial stress. Stress is a normalized variable, where the force applied is divided by the area over which the force is distributed. Axial stress is:

$$\sigma = \frac{F}{A}$$

where $\sigma$ is axial stress, F is force in N, A is the cross-sectional area in m². Axial stress ($\sigma$) is in units of pascal (Pa), where 1 Pa = 1 N/m².

## Shear Stress

In Figure 14.1, the force for both torsion and bending loads is applied tangential or parallel to the long axis of the material. When the force is applied tangential or parallel to the long axis of the material, the internal stress created is termed shear stress ($\tau$), denoted by the Greek letter tau ($\tau$). The formula to calculate sheer stress is:

$$\tau = \frac{F}{A}$$

where $\tau$ is the shear stress, F is force in N, A is the cross-sectional area in m². Shear stress ($\tau$) is in units of pascal (Pa), where 1 Pa = 1 N/m².

## Strain

When a load is applied to a material, it will undergo a change in shape. This change in shape or deformation of the material is described by the strain ($\varepsilon$). Strain can be measured in absolute or relative units, see Figure 14.2. Strain is usually given in relative units or percent change. In Figure 14.2, the compression force causes 10 mm of absolute strain. The relative strain for this example is the absolute strain, 10 mm, divided by initial dimension, 100 mm, which results in a relative strain of 0.1 or 10%. Relative strain is a normalized variable.

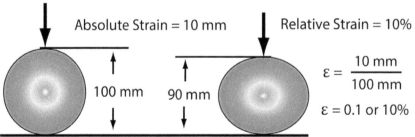

**Figure 14.2** Absolute and relative strain.

## STRESS-STRAIN RELATIONSHIP

The relative strength of a biological material can be quantified by graphing its stress-strain relationship. There is a direct relationship between the strength of a material and its stress–strain curve. Figure 14.3 shows an example of a stress–strain curve for a soft metal, such as copper. The stress-strain graph in Figure 14.3 illustrates the yield point, failure point, and stiffness for a material.

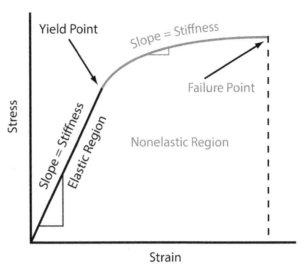

**Figure 14.3** Stress – strain graph for a ductile material showing elastic region, yield point and plastic region.

The maximum amount of deformation that a material can withstand while still being able to return to its original shape is defined as the **yield point**. When the material is loaded beyond the yield point, permanent damage occurs to the material and it can no longer return to its original shape. If the application of force is continued, the material will eventually break into two or more pieces, the **failure point** has been reached. The region of the stress – strain curve up to and including the yield point is termed the **elastic region**. In the elastic region, the material will deform during loading and return to its original shape when the load is removed. The region of the curve beyond the yield point is termed the **plastic region** or **non-elastic region**. In the plastic region, the material will deform during loading, when the load is removed the material does not return to its original shape. A plastic deformation results in permanent change of the material. If the magnitude of loading is just above the yield point, the permanent deformation may be microscopic and imperceptible.

The **stiffness** of a material is defined as the slope of its stress – strain curve. Stiffness is the ratio of stress to strain ($\sigma/\varepsilon$). In other words, stiffness is stress divided by strain. **Compliance** is the inverse of stiffness, so a material that deforms easily is said to be compliant and a material that is resistant to deformation is said to be stiff.

### Young's Modulus

During axial loading, many materials show a linear stress–strain response in the elastic region of the stress–strain curve. Young's modulus, also called the modulus of elasticity, describes the stiffness of a material in the elastic region during axial loading. If the material is uniform, such as a solid copper or aluminum rod, Young's modulus is essentially constant.

Uniform materials are known as isotropic materials. **Isotropic** is defined as having identical mechanical properties in all directions. Some materials do not exhibit a consistent mechanical response when the direction of loading is changed; these materials are said to be anisotropic. **Anisotropic** materials are direction–dependent. Impurities within the material or the direction of fibers within the material can produce an anisotropic mechanical response. Young's modulus is not constant for an anisotropic material.

## Stress-Strain Curves for Different Types of Materials

The stress–strain graph shown in Figure 14.4 depicts the typical mechanical response for loading of a ductile, brittle, and elastic material. Examples of ductile materials include bones and soft metals like copper or aluminum. A ductile material is stiffer below its yield point and more compliant above the yield point. The stress–strain curve for the elastic material indicates that elastic materials are initially compliant and become progressively stiffer as the load is increased. This mechanical behavior is clearly felt when stretching a rubber band, as it is initially very easy to deform, but becomes much stiffer as the elongation increases. Elastic materials show a linear stress–strain response in the elastic region of the curve. Some materials, such as a piece of chalk or a thin glass rod, will exhibit the stress–strain curve for brittle materials, as shown in Figure 14.4. When a brittle material is subjected to loading, it will fail (break) with little deformation. The yield point and the failure point are the same for a brittle material, thus it is not possible to bend a piece of chalk and produce a small crack in the chalk without breaking it.

## Viscoelastic Materials

Biological materials such as bones, ligament, tendons, and cartilage are elastic materials that contain fluid. Elastic materials containing fluid will exhibit an elastic (spring-like) and viscous (fluid-like) response when deformed. Materials exhibiting both viscous and elastic responses are defined as **viscoelastic**. A spring is an example of an elastic material. When deformed, it quickly returns to its original position when the load is removed. Molasses, syrup, and honey are examples of viscous materials. The thickness of the fluid is quantified by its vis-

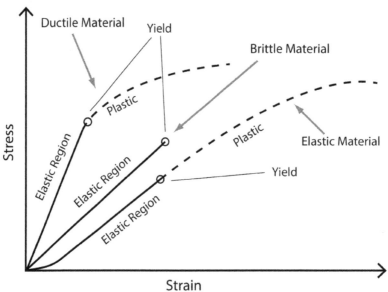

**Figure 14.4** Stress – strain graphs for ductile, brittle, and elastic materials.

cosity. **Viscosity** is a measure of the internal resistance of a fluid to flow. When a viscous fluid like molasses, is deformed it slowly returns to its original position after the load is removed. We refer to this slow return to the initial position as a **time-dependent response**. The response of a viscous material to deformation is affected by velocity and temperature. If you heat up the molasses, it will return to its original position quicker after deformation. The **velocity-dependent response** of viscous materials is also important to our understanding of the response of a biological material to loading. Viscous materials are stiffer when loaded fast and less stiff when loaded slowly. An example of the velocity-dependent response can be felt by moving your hand in water. First, move you hand slowly and feel the resistance to motion. Then move your hand as fast as possible, and again feel the resistance to motion. It should be clear that it is more difficult to move your hand in the fluid at a high velocity than it is at a low velocity. This is the velocity-dependent effect of viscous materials.

When loaded, a viscoelastic material will exhibit the following properties: creep, force relaxation, hysteresis, velocity-dependent response, time-dependent response, and finally a temperature dependent response. All of these properties are attributed to the viscosity, or viscous part of the material. Engineers use two elements to model the response of a viscoelastic material to loading: a spring and a dashpot (Figure 14.5). The spring models the elastic response and the dashpot models the viscous response. A shock absorber is a good example of a viscoelastic system. Most shock absorbers have a spring that is connected to a piston in a cylinder. The cylinder usually contains a fluid or gas that causes the shock absorber to dissipate or absorb energy.

The elastic response of the spring is linear and the force necessary to deform the spring is described by Hooke's law:

$$F = -kx$$

where $F$ is the force to compress or extend the spring in N, $x$ is the amount of compression or extension of the spring in m, and $k$ is the spring constant or stiffness of the spring in N/m. The spring is assumed to be an ideal spring. An ideal spring is defined as a massless spring that returns all of the energy stored in the spring during deformation.

Figure 14.6 depicts the mechanical response of an idealized (theoretical) spring. When the spring in the viscoelastic model above (Figure 14.5) is loaded, it will deform in a linear fashion directly related to the applied force and the stiffness of the spring. When the load is removed, the spring returns all of the elastic energy that was stored in the spring during deformation. As shown in Figure 14.6, when a spring is loaded and unloaded, its

Viscoelastic Model

Spring models elasticity

Dashpot models viscosity

**Figure 14.5** Example of a model used to describe the mechanical response of a viscoelastic material to loading.

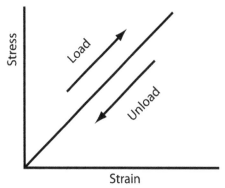

Stress

Load

Unload

Strain

**Figure 14.6** Stress–strain curve for an idealized spring.

stress–strain curve traces along the same line for the load and unload phase, suggesting that no energy is lost during deformation.

Unlike the spring, a viscoelastic material will absorb some energy during deformation. Due to this energy absorption, the unload curve for a viscoelastic material is different than the load curve (Figure 14.7). In Figure 14.7, the amount of **energy absorbed** by the material is equal to the area between the load and unload curves of the stress-strain graph. This area of energy absorption between the load and unload curves is called a **hysteresis**. The area underneath the unload region of the stress – strain curve represents the **energy returned** by the material (shaded region of Figure 14.7).

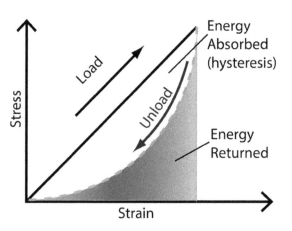

**Figure 14.7** Stress – strain curve for a viscoelastic material.

### Elastic Response as a Function of Time

A mass-spring system can be used to describe the effect of a spring on the motion of a body. Figure 14.8 shows the free body diagram for a mass-spring system. The spring in Figure 14.8 is considered as an ideal spring, as it applies an upward force to resist the downward pull of the weight. The spring force (Fs) is defined by Hooke's law, where k is the stiffness of the spring and x is amount of compression or elongation of the spring. The acceleration ($a_y$) of the block with mass *m* is defined as follows:

$$\Sigma Fy = ma_y$$
$$F_s + mg = ma_y$$
$$kx + mg = ma_y$$

The position of the weight as a function of time is determined by solving the following equation.

$$y(t) = kx(t) + mg + ma_y(t)$$

where *y* is position in m, *k* is the stiffness of the spring in N/m, *x* is amount of compression or elongation of the spring in m, $a_y$ is the acceleration of the weight in m/s², *t* is in s, and *m* is mass in kg. Figure 14.9 shows the motion of the mass-spring system as a function of time. At time t = 0, the spring is held in equilibrium (the spring is neither stretched nor compressed). The weight is suddenly released and it stretches the spring as it falls. The graph at the bottom of Figure 14.9 shows the result-

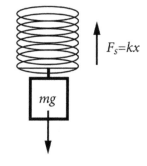

**Figure 14.8** A spring - mass system.

ing change in position of the weight as a function of time. The spring will eventually bring the weight to a momentary rest as the spring reaches its maximum stretch. For the spring shown in Figure 14.9, the first point of maximum stretch for the spring occurs at t = 0.26 s and the spring has been stretched 0.14 m below the equilibrium position. The elastic energy stored in the spring during the downward phase will now cause a restoration force that raises the weight back up to the equilibrium position. At t = 0.53 s, the weight has now returned to the equilibrium position. An ideal spring will continue to oscillate between +0.14 m and 0.0 m.

## Viscous Damping Force

In biological tissues, such as bones, cartilage, and tendons, the materials resistance to deformation exhibits damping, which is caused by fluid (air, water, synovial fluid) in the material. A damper is a device that attenuates, deadens, or restrains motion, or in other words, the damper absorbs kinetic energy. A piston in a sealed cylinder is an example of a damper. The damping effect can be modulated using fluids of different thickness (viscosity) to attenuate motion. The magnitude of damping is also affected by the velocity of motion. Damping increases with increasing velocity. The damping force ($F_d$) is computed as follows:

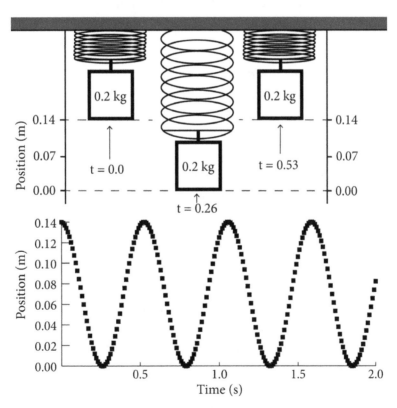

**Figure 14.9** Position of a mass-spring system as a function of time. At t = 0.0, the spring is released from a position of equilibrium. The 0.2 kg mass elongates the spring, storing energy in the spring. At t = 0.26, the spring stops moving downward. The spring will continue to oscillate between peak height of 0.14 m and the minumum height of 0.0 m.

$$F_d = cv$$

where $F_d$ is the damping force in N, $c$ is the coefficient of damping with units (Ns/m), and $v$ is the velocity with units (m/s). Using our example of a piston in a cylinder, the damping force might be relatively low if the cylinder contains a thin oil and the velocity of the piston is also low. Increasing the velocity of the piston will directly increase the damping force. The highest possible damping force for our piston example can be attained by the combination of using a very heavy grease in the cylinder and maximizing the velocity of the piston. Another example of a damping response can be seen by dropping a basketball and observing the rebound height of each bounce. Each bounce returns to a lower height than the previous bounce. In this example, the air in the basketball is the fluid causing the damping. Increasing the air pressure reduces the damping, while lowering the air pressure increases the damping.

### Viscous Damping Response as a Function of Time

The free body diagram of a mass–spring–damper system shown in Figure 14.10 is an example of a viscoelastic system. The spring and weight are now connected in series to a damper (also called a dashpot). The spring determines the elastic response and the damper determines the damping response of the system. The acceleration ($a_y$) of the weight is now defined as follows:

$$\Sigma Fy = ma_y$$
$$F_s + F_d + mg = ma_y$$
$$kx + cv + mg = ma_y$$

where $c$ is the coefficient of damping in Ns/m, $v$ is the velocity in m/s, $k$ is the stiffness of the spring in N/m, and $x$ is amount of compression or elongation of the spring in m. The position of the weight as a function of time for the mass-spring-damper system is determined by solving the following equation:

$$y(t) = kx(t) + cv(t) + mg + ma_y(t)$$

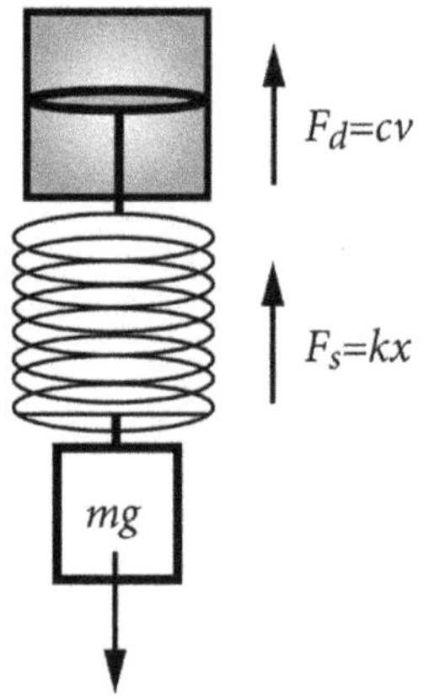

Figure 14.10 A mass - spring - damper system.

where $y$ is position in m, $k$ is the stiffness of the spring in N/m, $c$ is the coefficient of damping in Ns/m, $x$ is amount of compression or elongation of the spring in m, $a_y$ is the acceleration of the weight in m/s², $t$ is in s, and $m$ is mass in kg.

Figure 14.11 shows the motion of the mass-spring-damper system as a function of time. At time t = 0, the spring is held in equilibrium (the spring is neither stretched nor compressed). The weight is suddenly released and it stretches the spring as it falls, storing elastic energy in the spring. The inclusion of the damper in the system restricts the motion of the weight by opposing the motion of the weight proportionally to product of the damping coefficient ($c$) times the velocity ($v$). As a result of the damper restricting motion, the weight only falls to a minimum position of 0.02 m at t = 0.26 seconds. From this position of maximum

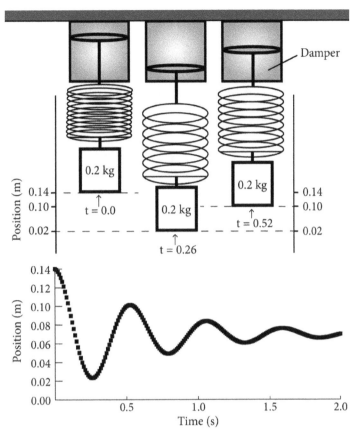

**Figure 14.11** Position of a mass-spring-damper system as a function of time. At t = 0.0, the spring is released from a position of equilibrium. The 0.2 kg mass elongates the spring, storing energy in the spring. The damper opposes the motion of the mass, absorbing kinetic energy. The resulting position-time curve is called a damped oscillation.

stretch, the restoring force in the spring will cause the weight to again move upward, reaching a position of 0.1 m at t = 0.52 s. Unlike the spring only system shown in Figure 14.9, each oscillation of the spring-mass-damper system is less than the previous oscillation. Eventually, the weight will come to rest as the damper continues to dissipate energy on each oscillation. The graph at the bottom of Figure 14.11 shows the resulting change in position of the weight as a function of time for the spring - mass - damper system.

## Comparison of Mass-Spring to Spring-Mass-Damper Systems

The graph shown in Figure 14.12 compares the motion of the weight between the mass-spring system and the mass-spring-damper system. In the mass-spring system (dark shaded boxes), the weight continues to oscillate back and forth between 0.14 m and 0.0 m. In the mass–spring–damper system (light shaded boxes), each oscillation of the weight is less than the previous oscillation. The spacing of the boxes in Figure 14.12 can be used to estimate the velocity of the weight. When the boxes are close together, the weight is moving slowly, when the boxes are spaced farther apart the weight is moving faster. For the mass–spring system (Figure 14.12), the weight attains a maximal downward velocity of –0.83 m/s occurs at follow-

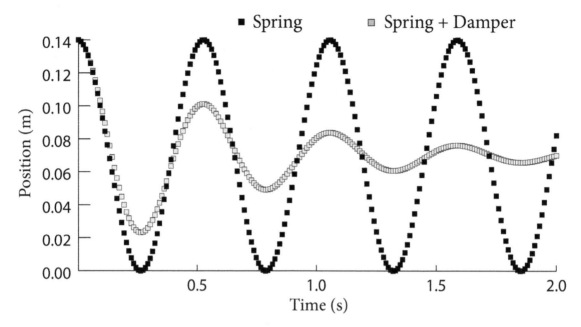

**Figure 14.12** Comparison of a mass-spring (black boxes) to a mass-spring-damper (gray boxes) as a function of time.

ing time points: t = 0.13, t = 0.66, t = 1.19, t = 1.72 s. Maximum upward velocity of +0.83 m/s, for the weight occurs at the following time points t = 0.40, t = 0.93, t = 1.46, t = 1.99 s. In the mass–spring–damper system, maximum downward velocity of –0.69 m/s occurs at t = 0.12 s and maximum upward velocity of +0.46 m/s occurs at t = 0.38 s.

### Creep Response of a Viscoelastic Material

The creep response for a viscoelastic material can be observed by applying a constant load to the material and measuring the deformation of the material over time. In Figure 14.13, a constant force of 30 lbs is applied to the material. When the load is applied, the material deforms quickly until the internal stress in the material reaches a relatively stable stiffness level (dashed line in Figure 14.13). After attaining this initial change in length, the material undergoes creep, where it slowly continues to deform over time (solid line). Creep is caused by viscous damping in the material. The creep response will be affected by the velocity of loading.

### Force Relaxation Response of a Viscoelastic Material

The force relaxation response of a viscoelastic material is shown in Figure 14.14. In force relaxation, the material is deformed to a fixed length and the change in force over time is measured. As shown in Figure 14.14, when the material is initially deformed, the force rises rapidly (A to B, dashed line). The material is stiff during this initial loading. If the material is held at the new position, the fluid in the material will settle and cause the internal resistance of the material to decrease. This decrease in stiffness is known as force relaxation (solid line from B to C).

The fluid in the viscoelastic material is responsible for the hysteresis, creep, and force relaxation. All viscoelastic materials will be stiffer when loaded quickly than they are when loaded slowly. This velocity-dependent response is attributed to the fluid in the material. At high rates of loading, the fluid is compressed as it attempts to flow from one region to another in the material. The stress–strain response of viscous fluids is also altered by temperature such that increasing the temperature lowers the stiffness and decreasing the temperature increases the stiffness

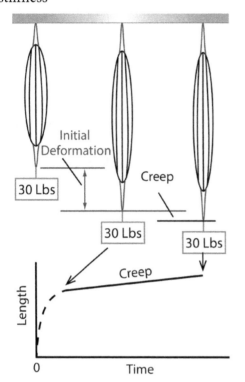

**Figure 14.13** Creep response for a viscoelastic material. The dashed line shows the intial deformation after the weight is released, and the solid shows the slower creep response.

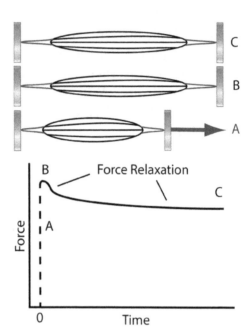

**Figure 14.14** Force relaxation response for a viscoelastic material. The dashed line shows the rapid rise in force when the muscle is pulled, and the solid line shows the slower force relaxation response.

## Review Questions

1. Define the following terms: stress, strain, stiffness, compliance.
2. What is the difference between absolute and relative strain?
3. Draw a picture of each of the following types of loads: compression, tension, shear, and torsion.
4. Draw a stress strain curve for a ductile, brittle, and elastic material. On each graph, identify the elastic region, yield point, and plastic region.
5. Define viscoelastic; explain the elastic response and the viscous response.
6. Describe the difference between a creep response and a force relaxation response when loading a viscoelastic material.

## SUGGESTED SUPPLEMENTAL SOURCES

Enoka, R. M. (2008). Neuromechanics of Human Movement. Champaign, Human Kinetics.

Franklin, K., P. Muir, et al. (2010). Introduction to Biological Physics for the Health and Life Sciences. Hoboken, John Wiley & Sons.

Griffiths, I. W. (2006). Principles of Biomechanics & Motion Analysis. Philadelphia, Lippincott Williams & Wilkins.

Hall, S. J. (2007). Basic Biomechanics. Boston, McGraw Hill.

Hamill, J. and K. M. Knutzen (2009). Biomechanical Basis of Human Movement. Philadelphia, Lippincott Williams & Wilkins.

McGinnis, P. M. (2013). Biomechanics of Sport and Exercise. Champaign, Human Kinetics.

Nordin, M. and V. H. Frankel (2001). Basic Biomechanics of the Skeletal System. Philadelphia, Lippincott Williams & Wilkins.

Whiting, W. C. and R. F. Zernicke (1998). Biomechanics of Musculoskeletal Injury. Champaign, Human Kinetics.

# Chapter 15

## EFFECTS OF EXERCISE ON BIOLOGICAL TISSUES

### LEARNING OUTCOMES

**After reading Chapter 15, you should be able to**

- Describe bone structure.
- Describe mechanical testing of biological materials.
- Explain the viscoelastic nature of bones, cartilage, ligaments, and tendons.
- Explain the relationship between type of load and fractures in bone.
- Describe bone modeling and remodeling.
- Explain the principles of healthy bones for future physical educators.
- Describe mechanical properties of cartilage, ligaments, and tendons.
- Explain the biomechanical descriptors of injury.

### BONE STRUCTURE

From an engineering perspective, the design of bone tissue is very impressive. Bone is said to be as strong as iron, yet as light as wood. Bone tissue derives this combination of strength and flexibility through the structural design of the cells that compose it and the mineral composition of bone tissue. Bone strength is not optimized by increasing the mass of the bone. Instead, bone strength is optimized by utilizing as little mass as needed and maximized by strategically altering the size, shape, and distribution of the mass. The load-bearing bones, such as the tibia and femur, are 2–4 times stronger than they need to be, based upon the peak stresses experienced during strenuous activity (Biewener, 1989). The exception to this is the skull, which is much more massive than required by typical loads. The skull must be over-designed to protect the brain from a monotonic failure (Rubin, et al., 1990).

### HISTOLOGICAL ORGANIZATION OF BONE

Cortical, or compact, bone is found on the outer edge of most bones and it forms the shafts of long bones, such as the tibia, femur, and humerus. Cortical bone is dense and highly mineralized, with a porosity of 5–15%. Mature bone is composed of the protein collagen and the minerals calcium and phosphate. The cortical bone matrix is primarily composed of these extracellular substances: collagen, calcium and phosphate. Osteoblasts deposit the bone matrix in thin, cylindrical layers, that form columns, which are oriented on the longitudinal axis of the bone. These cylindrical layers are called lamellae (see Figures 15.1). The lamellar layers of cortical bone are similar to a hollow cylinder and the bone tissue is laid down such that several lamellar layers are wrapped like a cylinder inside a cylinder to form a column that runs vertically down the long axis of a bone. A collection of these lamellar layers placed inside each other is referred to as an **osteon**, also called a **Haversian system**. The center of the osteon contains a hollow canal referred to as a Haversian canal, which usually contains a blood vessel and

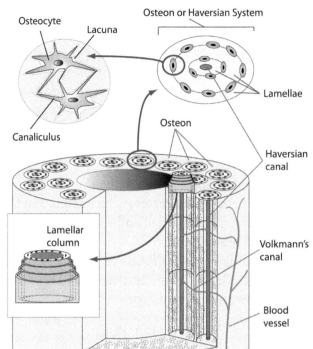

**Figure 15.1** Osteocytes in cortical bone are encased in lamellar layers. Several lamellar layers wrap around each other to form an osteon, also called a Haversian system. The Haversian canal in the center of the osteon contains blood vessels and nerve fibers. The Volkmann's canal forms interconnections between adjacent osteons.

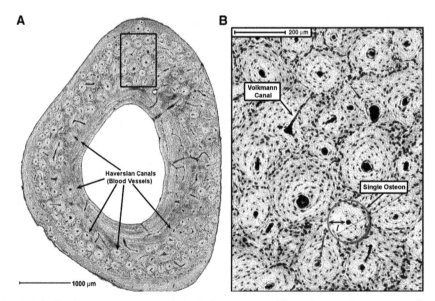

**Figure 15.2 A**. Midshaft of canine metacarpal. **B**. Enlargement of boxed area. Osteons are distributed in the majority of the cortex area (~ 80%). The Haversian and Volkmann's canals are well-connected with the marrow cavity.

From Gardinier JD, et al. In situ permeability measurement of the mammalian lacunar-canalicular system. *Bone*, 46: 1075-1081, 2010. Used with permission from Elsevier.

a nerve fiber. The Haversian canals of osteons are interconnected with horizontally-oriented canals known as Volkmann's canals. The Volkmann's canals establish communication between adjacent osteons, the free bone surface, and the marrow cavity. Collectively, this system of canals (Haversian and Volkmann's) provide nerve and blood supply for osteocytes.

During the formation of each layer (lamellae), the osteoblasts become encased in the bone matrix. These encased osteoblasts then form into mature osteocytes. Each osteocyte is enclosed inside a lacuna, which contains fluid. The dendrite-like projections extending outward from each osteocyte are called canaliculi. The canaliculi form connections with neighboring osteocytes and provide a means for communication and the exchange of substances by diffusion.

Cancellous bone, also called spongy or trabecular bone, has a porosity that varies from 30–90%. Cancellous bone is frequently found in the ends of the long bones and the interior of vertebrae. Cancellous bone is more compliant (less stiff) than cortical bone. Cancellous, or

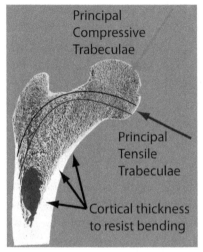

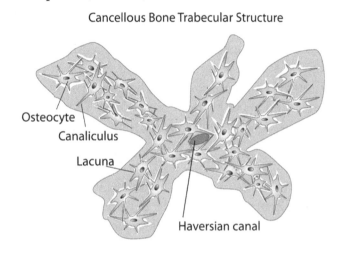

**Figure 15.3** Trabecular alignment matches principal loading patterns. Femoral image courtesy of Bartleby. com, Inc.

**Figure 15.4** Osteocytes in trabecular bone form in lamellar layers. The lamellar layers form trabecular plates and rods that align in the direction of loading. As in cortical tissue, the canaliculi form connections with neighboring osteocytes and provide a means for communication.

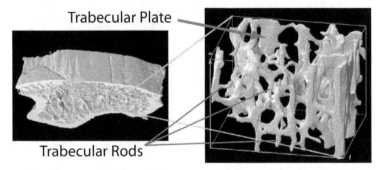

**Figure 15.5** CT image of a rat's proximal tibia (left). Image on the right shows trabecular plates and rods.

From Stölken JS, Kinney JH. On the importance of geometric nonlinearity in finite-element simulations of trabecular bone failure. *Bone*, 33: 494-504, 2003. Used with permission from Elsevier.

trabecular, bone, when healthy, is very effective in attenuating joint forces by distributing the force over a large surface area and absorbing impact energy by deformation. The structural orientation of cancellous bone tissue is somewhat analogous to a cube of Swiss cheese. The lattice-like structures of cancellous bone are called trabeculae. The trabecular matrix serves a mechanical function similar to the trusses found on a bridge in that they provide strength for the material while minimizing the weight of the structure. The orientation of the trabecular matrix is aligned in the direction where the greatest stresses are applied to the bone tissue (Cowin, 1986; Fyhrie, et al., 1990). For example, in the femoral neck, the trabecular tissue is aligned in the direction of principal compression and tensile loading (see Figure 15.3). The trabecular matrix is deposited in lamellar layers, but rather than wrapping cylinders around cylinders as found in cortical tissue, the lamellae of trabecular bone are formed as a mesh of plates and rods (Liu, et al., 2009; Liu, et al., 2008; Liu, et al., 2006; Liu, et al., 2009; Mosekilde, 2000; Stauber, et al., 2006), see Figures 15.4 and 15.5. The majority of the plates are aligned to withstand the principal direction of loading and the trabecular rods serve as transverse connections between the trabecular plates (Liu, et al., 2008). Osteocytes in the trabecular matrix are encased in a lacuna they also have canaliculi projecting outward to form connections with neighboring osteocytes in the trabecular tissue. Cancellous bone is highly vascular and it frequently contains bone marrow. During the formation of lamellar layers of cancellous tissue, a blood vessel and nerve become encased in a Haversian canal within the trabecular plates and rods. These blood vessels provide nutrients to the trabecular tissue.

## OSTEOCYTES

Cortical bone tissue in humans contains 13,900–19,400 osteocytes per mm$^3$ and trabecular bone contains 13,000 osteocytes per mm$^3$ (Sissons, et al., 1977). Osteocytes form an interconnected network through their dendrites. Individual osteocytes are thought to communicate with each other and the bone surface lining cells via their dendritic connections (Hazenberg, et al., 2007). The dendrites, or canaliculi, form an intercellular communication network with neighboring osteocytes and with osteoblasts located on the bone surface. Figure 15.6 shows canaliculi forming this intercellular network.

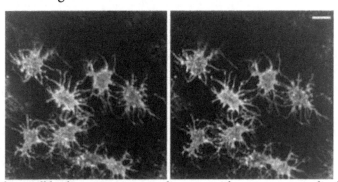

**Figure 15.6** Osteocytes have cell bodies ~15–20 μm in diameter with numerous canaliculi that form an intercellular communication network.

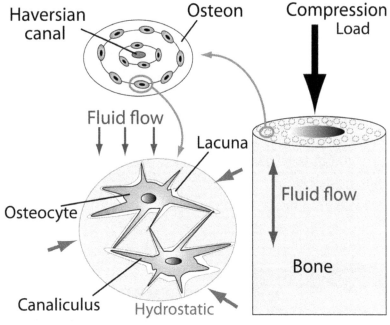

Adapted from Chen (2010) J Biom, 43:108-118.

**Figure 15.7** Osteocytes are embedded in osteons and arranged around a central Haversian canal. Mechanical loads on bone generate fluid flow in the canalicular space, eliciting biochemical responses to the loading in osteocytes.

From Chen H, et al. Boning up on Wolff's law: Mechanical regulation of the cells that make and maintain bone. *J Biomech*, 43: 108-118, 2010. Adapted with permission from Elsevier.

During bone loading, the osteocytes are thought to measure the strains induced by various loads. When a force is applied to bone tissue, it will undergo deformation. This deformation will cause fluid to move within the bone matrix, forcing fluid through the Haversian and Volkmann's canals. Figure 15.7 shows compression loading along the long axis of a bone such as the femur or tibia. The compression load will cause deformation of the bone tissue and movement of fluid. The osteocytes are thought to measure the strain induced by applied loads. Mechanical loads on bone generate fluid flow in the canalicular space, eliciting biochemical responses to the loading in osteocytes.

## MECHANICAL TESTING OF BONE

Forces or loads always cause strain in biological materials, even if that strain is imperceptible. Biomechanists often express strain in microstrain ($\mu\varepsilon$) units, where 1,000 microstrain in compression would shorten a bone by 0.1% of its original length, 10,000 microstrain would shorten it by 1% of that length, and 100,000 microstrain would shorten it by 10% of initial length. Table 15.1 gives microstrain units and their corresponding % change of length. The physiological strain range for the long bones is 400–3,000 $\mu\varepsilon$, with typical loading rarely above 1,000 microstrain. Bone begins to fail at 7,000 microstrain (0.7%) and the ultimate strength or failure point for bone under compression or tension is 25,000 $\mu\varepsilon$, which corresponds to

### Table 15.1 Strain Units & Physiological Strain Ranges for Long Bones

| Microstrain Units | | % Change in Length |
|---|---|---|
| 1,000 microstrain | 1,000 με | 0.1% |
| 10,000 microstrain | 10,000 με | 1.0% |
| 25,000 microstrain | 25,000 με | 2.5% |

| Physiological Strain Ranges for Long Bones | | |
|---|---|---|
| 400–3,000 microstrain | 400–3,000 με | 0.04–0.3% |
| Seldom above 1,000 microstrain | 1,000 με | 0.1% |

Figure 15.8 Example of a computer-controlled machine used to mechanically test biological materials.

Figure 15.9 Example of three-point bending to determine the mechanical properties of a rat femur.

Photo courtesy of Instron®.

Photo courtesy of Instron®.

a change in length of 2.5%. As result, bone fails at strains of 25,000 microstrain, where the length changes from 100% to 97.5% under compression and from 100% to 102.5% under tension (Frost, 1997).

Mechanical testing of biological materials (bone, tendons, ligaments, cartilage) is usually done with material testing system (Figure 15.8 for an example). The material to be tested is mounted on the machine and then the machine is programmed to deform the material and subsequently measure its stress–strain response to the imposed load. The amount of deformation and the rate of loading can be varied within the limitations of the machine. These machines can perform compression, tension, torsion, and bending tests. Figure 15.9 shows an example of using a material testing system to determine the bending properties of a rat femur.

The stress–strain curves for cortical and cancellous bone are shown in Figure 15.10. Bone is a ductile material and as a result, it is stiffer below the yield point. Since cortical tissue is highly mineralized and is considerably denser than cancellous tissue, it has a steeper stress–strain curve. Cortical bone yields at a higher force and lower strain than cancellous bone.

## VISCOELASTIC NATURE OF BONE

Cortical and cancellous bone tissues both have fluid within the tissues. As illustrated in Figure 15.7, applying a load to bone tissue causes increased hydrostatic pressure within the tissue. Bone, like all viscoelastic materials, is velocity–dependent. Therefore, when loaded at a fast rate, the tissue is stiffer when compared to loading at a slower rate. Figure 15.11 shows the effect of the rate of loading on the stress-strain response in bone tissue. There is considerably

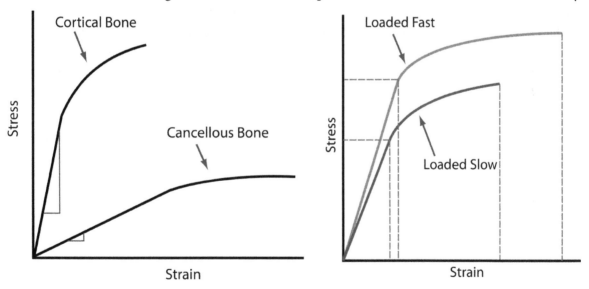

**Figure 15.10** Stress-strain curves for cortical and cancellous bone. Cortical bone is stiffer, it fails at a lower strain and higher stress than does the more compliant cancellous tissue.

**Figure 15.11** Since bone tissue is viscoelastic, it is stiffer when loaded at a fast rate than when loaded at a slower rate.

less fluid in cortical bone than there is in cancellous bone tissue. Like all viscoelastic tissues, bone is stronger when loaded at a fast rate than when loaded at a slow rate.

## RELATIONSHIP BETWEEN TYPE OF LOAD AND FRACTURE IN BONE

Figures 15.12 and 15.13 show the relationship between the type of load applied to bone and the resulting fracture. When bone tissue fails under tension load, the resulting fracture is called an avulsion fracture. In an avulsion fracture, a piece of bone tissue is torn away (shown on the right in Figure 15.12). Muscles and ligaments tend to apply very large tension forces that often result in avulsion factures. When bone tissue fails under torsion loads, the resulting fracture is spiral fracture. Baseball players occasionally get spiral fractures of the humerus. During the reversal phase of the throwing motion, one end of the humerus rotates internally while the other end rotates externally, causing a torsion load. Gymnasts and skaters also get spiral fractures if they attempt to complete a twisting motion after they land from a flight phase. Bending forces produce a triangular or trapezoidal fracture (see Figure 15.13). The triangular or trapezoidal fracture occurs due to compression forces on the inside of the bend and tension forces on the outside of the bend. Both bending and torsion loads result in lines of shear where there are parallel opposing forces within the tissue.

## BONE ADAPTATION

Bone tissue is remarkably resilient and adaptable. There is a tendency to think of bone as inert and resistant to change, when in fact, bone is very dynamic tissue. Introductory exercise science students can all readily quote Dr. Julius Wolff in what is commonly known as Wolff's Law: "Every change in the function of a bone is followed by certain definite changes in its internal architecture and its external conformation." (Brand, 2010; Chen, et al., 2010; Wolff, 1870). We now understand that bone responds immediately to exercise that causes strains above or below the level expected. Current evidence suggests that small cracks, microdamage,

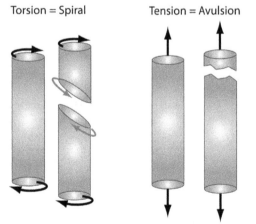

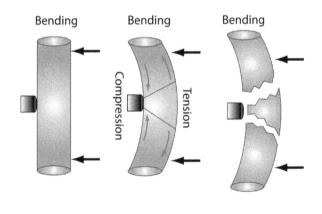

**Figure 15.12** A torsion, or twisting, force results in a spiral fracture to bone tissue (left). Tension forces often cause an avulsion fracture in bone tissue (right).

**Figure 15.13** Bending of bone tissue causes compression forces on the inner surface and tension forces on the outer surface of the bend. Bending forces cause triangular or trapezoidal fractures to bone tissue.

in the osteocyte network may appear as soon as 24 h post-loading which then cause the osteocytes to initiate an adaptive remodeling response. Microdamage in the osteocyte network is thought to disrupt communication and fluid flow in the local region. Disruption of inter-osteocyte communication triggers bone remodeling (Allen, et al., 2008).

The adaptation of bone is controlled by cellular mechanisms. The building of new bone during growth is referred to as modeling, whereas rebuilding of bone tissue as a consequence of injury or variations in activity is referred to as remodeling. During growth, bone modeling produces a change in the size and shape of bone when new bone is deposited without resorption of existing bone tissue. Bone remodeling, on the other hand, follows a sequence consisting of activation, resorption, and formation. Bone remodeling begins with resorption of existing bone tissue by osteoclasts, which is then followed by osteoblast activity to form new bone tissue. Remodeling of bone is accomplished by the interaction of four types of cells, collectively called basic multicellular units (BMUs). BMUs include osteocytes, osteoclasts, osteoblasts, and bone lining cells (Frost, 1987). Bone lining cells, osteoblasts, and osteoclasts are located on the bone surfaces, while the osteocytes are located in the bone matrix.

Osteocytes are thought to be responsible for measuring the strains induced in bone and initiating an osteogenic response if the strains are above or below typical strain ranges of 400–1,000 microstrain. In most animals, peak functional strains range from less than 1,000 microstrain during walking, to 2,000–3,200 microstrain for vigorous activities (Fritton, et al., 2000). The **objective of adaptive remodeling** is to maintain the bone cell structural alignment and mineral density such that typical loads produce strains in the range of 400–1,000 microstrain. When typical loads result in less than 400 microstrain the osteocytes would initiate a response to cause bone tissue resorption, and when typical loads are consistently greater than 1,000 microstrain osteocytes would initiate a response that would lead to new bone being laid down to stiffen the tissue in the area experiencing above normal strain. In the default state, when typical loads result in strains of 400 – 1000 microstrain, it is believed that osteocytes inhibit a remodeling response. Higher magnitude of loading which results in strains greater than (2000 – 3200 $\mu\varepsilon$) have been shown to cause microcracks in the osteocyte network that appear as soon as 24 hours after a bout of loading. Small cracks in the canaliculi disrupts the normal osteocyte inhibition of osteoclasts. The removal of osteoclast inhibition is called **disinhibition**. Disinhibition triggers the bone remodeling within minutes of excessive loading (Ehrlich, et al., 2002). A biochemical signal to the lining cells at the surface of the bone tissue initiates a sequence of events that lead to eventual bone remodeling (Allen & Burr, 2008; Rochefort, et al., 2010). Osteocyte with broken canaliculi undergo **apoptosis**, a form of cell-programmed death that causes the affected cells to disintegrate.

Figure 15.14 depicts the steps in bone remodeling. In step 1, the quiescent bone surface is displayed showing healthy osteocytes imbedded in the bone matrix and lining cells on the bone surface. Between steps 1 and 2, a force that exceeds the structural integrity of the bone tissue is applied and local strain causes a microcrack to occur in the bone matrix. Osteocytes whose canaliculi were affected by the disruption of communication initiate the bone remodeling response by removing the normal inhibition of osteoclastic activity. Disinhibition results in the transmission of a biochemical message to the bone surface that activates the os-

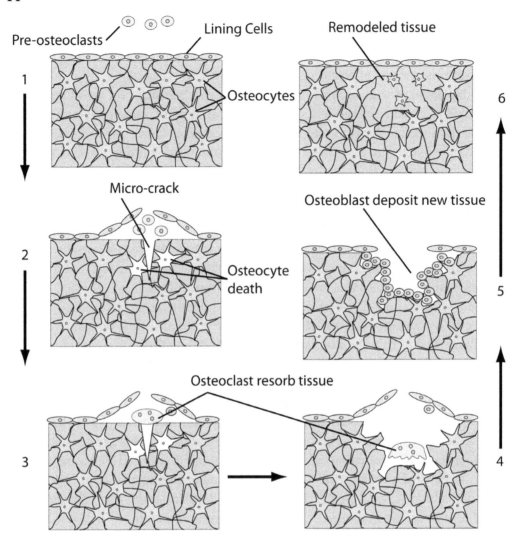

**Figure 15.14** Bone remodeling 1–6. 1) Quiescent bone surface with healthy osteocytes imbedded in the bone matrix. 2) Excessive loading causes microcrack. 3) Pre-osteoclast migrate to affected area and mature. 4) Osteoclast resorb affected cells. 5) Osteoblast deposit new tissue. 6) Osteoblast mature into osteocytes, some turn into new bone lining cells. The newly formed bone matrix will accumulate minerals and increase bone density over a 3 year period.

teoclasts. The lining cells in the affected area open up allowing osteoclasts to enter the area of tissue with the microdamage upon receiving the biochemical message from the osteocytes. In steps 3 and 4, the pre-osteoclasts mature and then begin excavating a cavity in the affected area. The affected osteocytes undergo apoptosis. The bone resorption by osteoclasts may take about two weeks. Between steps 4 and 5, the osteoclasts undergo apoptosis. In step 5, the osteoblast lay down new bone tissue. Between steps 5 and 6, as the new bone matrix fills in some of the osteoblast turn into osteocytes, some turn into lining cells and the rest undergo apoptosis. In step 6, the osteocytes in the newly formed bone matrix establish dendrite networks with neighboring osteocytes and the bone lining cells. The new bone matrix will accumulate minerals and increase bone density over a three year period.

Adaptive remodeling in bone is complex and situation-dependent. For example, static strains do not engender adaptive remodeling, whereas high-frequency impact loading induces a greater adaptive remodeling response than does low frequency loading (Ehrlich & Lanyon, 2002; Lanyon, et al., 1984). Current evidence suggest that an optimal osteogenic response to exercise would be 36 loading events with 4–8 hours rest between exercise bouts (Burr, et al, 2002). Burr et al., (2002) summarized several studies that indicate that more is not necessarily better when performing exercise to stimulate bone growth, as the osteogenic response saturates at 36 cycles of loading and any additional loading has little effect.

## Detraining and Bone Remodeling

Prolonged decreases in functional loading in long bones of adults' results in a net decrease in bone tissue, as the lack of functional loading causes bone resorption to outpace bone growth (Rubin, et al., 1996; Vico, et al., 2000). The mineral content and architectural structure of bone must be adjusted so that typical loads induce adaptive remodeling while still preserving a sufficient safety factor to withstand atypical or unexpected loads. As stated above, the objective of bone remodeling is to design the cell orientation and mineral content so that typical daily loads result in strains of 400–1,000 microstrain. A prolonged decrease would then lead to resorption of bone tissue so that typical daily loads cause 400–1,000 microstrain.

## Why are Bones Curved?

Most long bones of the body, such as the femur, have a natural curvature. A purely straight column bone would be stronger under compression than would a curved bone. Therefore, there must be an evolutionary advantage for curved bones. It turns out that there are

two clear advantages of curved bones when compared to a purely straight column bone. Figure 15.15 shows a typical curved bone. The first advantage of a curved bone is that when loaded in the longitudinal direction, the bone will deform in a predictable direction. The second advantage of a curved bone is that it will bend more when the load is increased. The curved bone in Figure 15.15 bends more when the 2,000 N load is applied than it does with the 1,000 N load. Recall from the discussion above on the osteogenic stimulus that it is the amount of strain that determines the stimulus for adaptive remodeling. Thus, curved bones have an enhanced ability to differentiate between loading conditions and then alter the osteogenic stimulus according to the magnitude of strain (Lanyon, 1987).

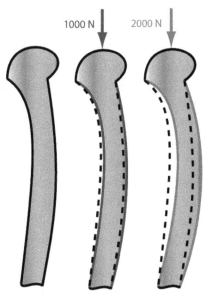

**Figure 15.15** A curved bone deforms in a predictable direction and deforms more under increased axial loading.

## DEGENERATIVE CHANGES IN BONE DUE TO AGING

The adaptive remodeling process described above is a continuous and dynamic process where the balance between bone architecture and mechanical loading determine the osteogenic response. Bone is constantly being resorbed and new bone is formed in its place. Bone mass usually peaks between ages 25–35. After age 35, women gradually lose bone mass. For men, this decline usually occurs after age 45. An individual's bone mineral content is dependent upon the peak bone mass achieved during youth and the subsequent rate of bone loss after age 35 for women and age 45 for men. In some individuals, the normal rate of bone loss after the third decade of life is steeper, causing them to develop osteoporosis. Osteoporosis is a skeletal disorder that is characterized by reduced bone strength. Individuals with osteoporosis have an elevated risk of fracture when compared to age-matched controls. Women typically experience a rapid phase of bone loss that begins before the cessation of menses and continues for up to five years post menopause. A decrease in the concentration of estrogen is thought to be the primary factor leading to osteoporosis in women. Other risk factors that may contribute to the development of osteoporosis include family history, low vitamin D and calcium intake, low body weight, and advanced age (National Osteoporosis Foundation, 2003). Osteoporosis is preventable and treatable when diagnosed early. Unfortunately, there are no warning signs until a fracture occurs. The National Osteoporosis Foundation recommends that all women be evaluated for osteoporosis, after menopause, by a having a bone scan to determine bone mineral density.

**Dual-energy x-ray absorptiometry (DEXA)**, which uses an enhanced form of x-ray, is the preferred method to determine bone mineral density (BMD). DEXA scans are used to diagnose osteoporosis or to asses an individual's risk for developing a fracture. Figure 15.16 shows a DEXA machine and a sample output displaying femoral neck bone mineral density. DEXA machines determine bone mineral density which is reported in $gm/cm^2$ and also converted to two normalized scales, T score and Z score. The T score compares the individual's BMD to that of a young adult of the same gender with peak bone mass. A T score above −1 is considered normal, a T score between −1 and −2.5 is classified as osteopenia (low bone mass), and a T score below −2.5 is defined as osteoporosis. A Z score compares the individual's BMD with that of other individuals of similar age, size, and gender. A Z score of less than −1 indicates that the individual is in the lowest 25% of his or her age group, and a Z score of less than −2 indicates the individual is in the lowest 2.5%.

Bone strength and fracture resistance is a function of bone density and architecture. It is possible for a bone to lose density without adversely compromising bone strength as long as the trabecular structure is aligned in the principal direction of loading. Unfortunately the normal aging process typically results in a gradual deterioration of both bone density and architectural structural alignment. Figure 15.17 shows vertebral bodies for individual (A) and an elderly individual (B) from a study by Mosekilde (2000). Mosekilde analyzed age-related changes in vertebral bone using histomorphometry, structural analyses, scanning electron microscopy, and biomechanical testing. In young individuals, Mosekilde found that trabecular bone volume was 15–20% of total bone volume, cortical thickness was 400–500 μm and axial load bearing capacity was 1,000–1,200 kg. For individuals between the ages of 70–80, trabecular bone volume was 8–12% of total bone volume, cortical thickness was 200–300 μm and axial load bearing capacity was 150–120 kg. For individuals with osteoporosis, trabecular bone volume was 4–8% of total bone volume, cortical thickness was 120–150 μm, and axial load bearing capacity was 60–150 kg. These changes in bone density and structure in both the elderly and individuals with osteoporosis illustrate how changes in bone strength can result in an increase fracture risk associated with a fall.

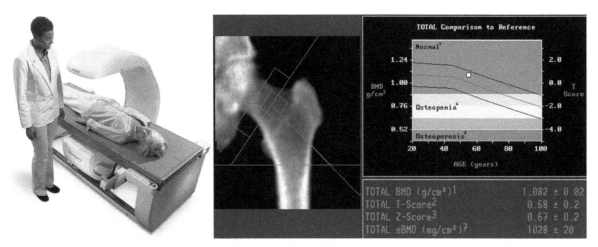

**Figure 15.16** A DEXA machine, shown on the left, is commonly used to measure bone mineral density to diagnose osteopenia or osteoporosis. The image shown on the right is a typical femoral neck scan. The scan results give BMD, T score, and Z score for the femoral neck.

Photo Courtesy of Hologic, Inc

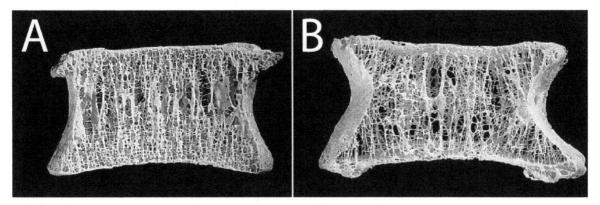

**Figure 15.17 A.** Vertebral body from a young individual. The trabecular matrix is dense and the architectural alignment supports axial loads. **B.** Vertebral body from an elderly individual. The trabecular density is decreased and the support for axial loads is diminished.

From Mosekilde L. Age-related changes in bone mass, structure, and strength-effects of loading. *Z Rheumatol*, 59 (Suppl 1):1-9, 2000. Used with permission from Steinkopff Verlag.

## Physical Activity and Bone Health

The beneficial effects of high levels of physical activity in childhood and throughout adult life may serve as a protective buffer against developing osteoporosis. After attaining peak bone mass in their thirties, individuals who maintain high levels of physical activity tend to show a slower rate of bone loss later in life. The osteogenic effect of exercise diminishes with advancing age. In addition, the gradual deterioration of articular cartilage in the knee and hip joints renders impacting activities such as running and jumping unsafe or even painful for elderly individuals. Thus, they are no longer able to execute the types of exercise shown to be helpful in minimizing bone loss. Fast walking, and vibration training have shown small, but promising, results indicating that age appropriate exercise training can improve bone mineral density. Bergström, et al. (2008) reported a small beneficial effect of fast walking (3 × 30 min/week) and physical training twice per week on total hip BMD in osteoporotic women ages 45 to 65 years. Rubin, et al. (2004) found a 2.17% beneficial effect in the femoral neck and a 1.5% beneficial effect in the spine in osteoporotic women who participated in a one year study, in which the subjects stood on a platform that induced vertical vibrations for 10 minutes per day for 7 days per week.

## PRINCIPLES OF HEALTHY BONE LOADING FOR PHYSICAL EDUCATORS

Physical educators are starting to incorporate the results of animal studies and physical intervention studies to maximize the osteogenic effect of exercise in physical education classes. In Vancouver, physical educators implemented a program called Action Schools! BC, which included an extra 15 minutes of simple activities for 5 days a week and a simple "bounce at the bell", where children performed 10 countermovement jumps 3 times a day (at each school bell). After eight months, bone mineral content of the proximal femur improved by approximately 2% in the exercise groups (McKay, et al., 2005).

Physical educators should follow these principles to implement healthy bone loading:

1. Ideally the children should perform 120–200 jumps per day, 5 days per week.
2. Different types of jumps should be performed each day or each week. For example: week 1 jumping rope, week 2 – hopping, etc.
3. Static loads are not as effective as dynamic loads.
4. Spreading the activity over 3–5 days is far more effective than doing the jumping/hopping activity all in one day.

## BIOMECHANICS OF CARTILAGE

In diarthrodial joints such as the knee, hip, elbow, and shoulder, the ends of the articulating bones are covered with a 1-6 mm layer of articular cartilage. Articular cartilage has two primary functions: attenuate force, and reduce friction. The joint space between the ends of the bones is filled with synovial fluid, which serves as a lubricant and a source of nutrients for articular cartilage. The combination of cartilage, bone, and synovial fluid provides diarthrodial joints with a low friction environment that minimizes wear of the articular surfaces. Despite the relative thinness (1-6 mm) of articular cartilage, it is a very effective force attenuator. Func-

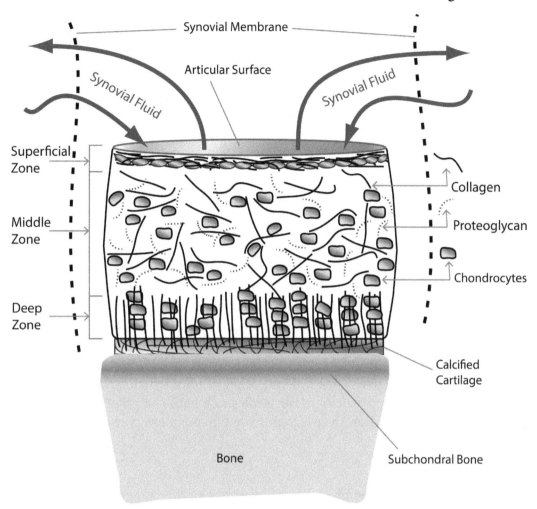

**Figure 15.18** Cell composition and structural arrangement of articular cartilage. Joint loading causes synovial fluid to flow out of the cartilage, pass thru the synovial membrane, and then flow back into unloaded regions of the cartilage.

tionally, articular cartilage can be thought of as a viscoelastic sponge that protects the underlying bone from harmful forces. Since cartilage has little or no blood supply, it must rely upon synovial fluid to provide nutrients for the cartilage and to remove waste products from the cartilage. Nutrients and waste products are delivered to the articular surface by synovial fluid and then subsequently transported into the articular cartilage via diffusion.

Articular cartilage is composed of chondrocytes, collagen fibers and proteoglycans. The cells in articular cartilage are called chondrocytes. Chondrocytes are the building blocks that produce and repair collagen fibers and proteoglycans (Lakin, et al., 2013). There are very few chondrocytes in articular cartilage. Chondrocytes compose 2–10% of the cartilage, collagen fibers compose 15–20%, proteoglycans compose 4–6%, and the remaining 70–80% of the cartilage is water. Joint loading causes deformation of the articular cartilage, which then alters fluid flow, hydrostatic pressure and osmotic pressure (Guilak, et al., 2006). These changes in pressure and fluid flow are assumed to play a role in signaling chondrocytes to take appropriate actions to maintain healthy cartilage. The strength of articular cartilage is largely determined by the amount and orientation of collagen fibers. Proteoglycans, are protein structures that are essentially the glue of articular cartilage, they keep the structure intact. The articular cartilage structures outside of the cells (collagen, proteoglycans, glycosaminoglycans, and chondroitin sulfate) are collectively referred to as the extracellular matrix. Finally, water serves two primary functions in articular cartilage: (1) when the joint is loaded, water builds up pressure to serve as a force attenuator, and (2) water is the delivery system for nutrients and the waste removal system for cartilage tissue.

Figure 15.18 depicts the chondrocyte, collagen fiber, and proteoglycan arrangement in articular cartilage. The articular cartilage is typically divided into three separate regions (superficial, middle, deep) based upon the collagen and chondrocyte alignment in the tissue (Rieppo, et al., 2009; Weiss, et al., 1968; Xia, 2008; Zheng, et al., 2009). As shown in Figure 15.18, chondrocytes in the superficial zone are aligned parallel to the articular surface and the cells are more flattened then the cells in the middle and deep zones. The collagen fibers in the superficial zone are also aligned parallel to the articular surface. They are densely packed and interconnected to give the superficial zone a rigid surface that is effective in dissipating or spreading out applied forces. In the middle zone, the chondrocytes and collagen fibers are randomly arranged, and in the deep zone, the chondrocytes are stacked in a vertical fashion (Lakin, et al., 2013). Collectively, this cell and collagen fiber structure behaves similar to a football helmet in that the top articular layer serves as a hard shell that spreads the applied force out and the middle and deep layer serves as two different density springs to attenuate the force.

## Synovial Fluid

Synovial fluid is produced by the synovial membrane (synovium). Synovial fluid is a thick, viscous fluid with the consistency of egg yolk. It has a high affinity to adhere to the articular surface and it forms a protective film layer on the exterior surface of the articular cartilage. The function of synovial fluid is to provide nutrients for the cartilage, lubricate the articular surface to reduce wear, and serve as a shock absorber to work with cartilage to attenuate loading forces across the joint (Schurz, et al., 1987). Synovial fluid is a special type of fluid referred to as a non-Newtonian fluid. The viscoelas-

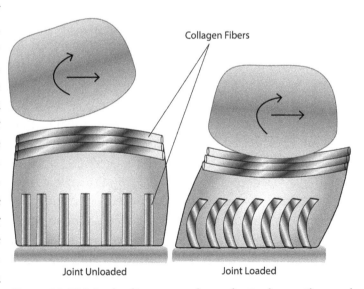

**Figure 15.19** Joint loading causes shear of articular cartilage and deformation of collagen fibers in the cartilage.

tic response of a non-Newtonian fluid is affected by the rate of shearing in the fluid. For synovial fluid, the viscoelastic behavior of the fluid is due to the amount of hyaluronic acid. When exposed to a low rate of shear force loading, hyaluronic acid has a high level of viscosity and a low level of elasticity. In contrast, when the rate of shear loading is high, viscosity is reduced, making the fluid thinner, and the elasticity is increased, enabling the fluid to absorb additional energy (Balazs, et al., 1993). This non-Newtonian response of hyaluronic acid gives synovial fluid a velocity-dependent response, where the fluid is a good lubricant at low rates of loading and a good shock absorber for high rates of loading.

## Effects of Loading Rate on Cartilage

The viscoelastic response of articular cartilage to loading is attributed to the tissue structure and composition, and the presence of a layer of synovial fluid on the articular surface. Articular cartilage can be described as a composite material that is fiber-reinforced (collagen), fluid-saturated (water), and porous. Proteoglycans in the cartilage give off a negative charge, which causes unloaded regions of articular cartilage retain water and become saturated (Maroudas, 1976). When saturated, the pressure within the cartilage matrix causes the cartilage to bulge and the collagen fibers on the articular surface are preloaded in tension. Figure 15.19 illustrates shearing of articular cartilage during joint loading and the effects of loading on collagen fibers. The image on the left in Figure 15.19 shows the collagen fibers on the articular surface bulging up in the direction of the joint space when the joint is unloaded. The bone above the joint surface is moving to the right and rotating in a counterclockwise direction relative to the articular cartilage. When the bone applies a downward and forward force on the articular cartilage, it will cause the cartilage to shear and compress as shown on the right in Figure 15.19. Viscoelastic force attenuation in articular cartilage can be attributed to: shearing and compression of the collagen fibers, forcing fluid out of the cartilage, and the non-Newtonian behavior of synovial fluid. Increasing the rate of loading will increase the elasticity of synovial fluid and the stiffness of the articular cartilage, as the fluid attempts to exit a limited number of pores in the articular surface. These properties give articular cartilage a rate-dependent response where the material is stiff when loaded at a fast rate and compliant when loaded at a slow rate.

For example, in walking, the rate of joint loading is low and therefore the cartilage-synovial fluid system is more compliant (attenuates less force). In contrast, when exposed to a high rate of loading, as in the takeoff for a long jump, the elasticity of synovial fluid increases and the cartilage-synovial fluid system is much stiffer (attenuates more force). In essence, the articular cartilage-synovial fluid system can be thought of as an adjustable shock absorber system. Under low rates of loading, it is not stiff, and it passes the load on to the bone tissue. Under high rates of loading ,the cartilage-synovial fluid system is stiff and it protects the bone from harmful high-frequency forces. It is remarkable that this adjustable shock absorber system is contained in only 1-6 mm of tissue.

## Healthy and Unhealthy Exercise for Articular Cartilage

When a load is applied to the joint, the water is forced out of the cartilage and it flows freely in and out of the synovial membrane. The synovial membrane is not a barrier to synovial fluid. The synovial membrane has a rich supply of blood vessels and it serves as the primary source of nutrients for the cartilage tissue. Exercises such as walking and running, where the joint alternates between being loaded and unloaded, are considered to be healthy for cartilage tissue. On the other hand, static loading of joint is unhealthy for cartilage tissue, as the synovial fluid is forced out of the cartilage but not allowed to flow back into the cartilage until the joint is unloaded. Individuals who stand for extended hours without moving during work may eventually cause permanent fatigue failure of the cartilage, causing it to flatten out. Long term high-impact loading may also lead to damage to articular cartilage. In addition, repetitively applying compression or torsion forces may deteriorate the area where the load is applied. For example, a right-handed golfer repetitively loads and applies torsion forces to a relatively narrow region of articular cartilage of the left knee. It is not too hard to imagine how this can cause permanent fatigue and damage to the underlying articular cartilage tissue.

### Causes of Degeneration in Articular Cartilage

Since articular cartilage is devoid of a direct blood supply, it has limited ability to maintain and repair itself. Degeneration of articular cartilage adversely affects its ability to lubricate the joint and attenuate forces during joint loading. In addition, degeneration of cartilage tissue leads to stiffening of the subchondral bone. Together, these alterations in articular cartilage and subchondral bone often lead to joint pain and potentially osteoarthritis. So what are the potential causes of wear in articular cartilage? Wear of articular cartilage is generally attributed to some combination of excessive impact loading, chronic static loading of the joint, chronic dynamic loading of a specific area, and finally, trauma from an injury that either tears the articular surface or a blunt force that disrupts the underlying cartilage matrix.

In the 1970's, many people started running to improve their cardiovascular fitness, but, in the 1970's, running shoes had little or no ability to attenuate impact forces and to control for excessive pronation. This led to many runners experiencing knee pain. As the running trend continued, orthopedic surgeons saw an increasing number of relatively young people with knee joints that were osteoarthritic. In 1973, Radin theorized that it was the repeated impacts from running that was leading to osteoarthritis in the runners. He developed an animal experiment to test his theory (Radin, et al., 1973). Figure 15.20 shows his exper-

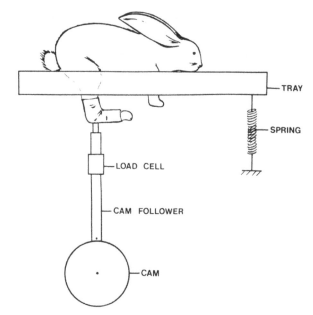

**Figure 15.20** The knee joints of adult rabbits were subjected to daily 1 hour intervals equal to 1 BW, at 60 cycles per min.

From Radin EL, et al., Response of joints to impact loading-III. *J Biomech*, 6:51-57, 1973. Used with permission from Elsevier.

imental setup. The rabbit's right leg was impacted with a force of 1 body weight that reached a peak within 50 ms, while the left leg was not impacted. The rabbits were exposed to one hour of impacting each day and then allowed to roam freely in their cage. Groups of animals were sacrificed by twos at two-day intervals for up to 36 days. The rabbits that experienced several days of impacting eventually developed a slight right-sided limp. Upon sacrifice, the animals' knees were analyzed anatomically and mechanically. The results showed that the cortical bone on the ends of the tibia and femur was thicker, the cancellous bone was brittle, and the cartilage was thin or non-existent. The cancellous bone contained numerous microcracks that when healed caused the cancellous bone to stiffen. The bone tissue stiffened first, followed by deterioration of the cartilage tissue. This study, and many others like it, led to the general consensus that excessive impact loading of joints is one factor that can lead to joint deterioration and for some individuals, the need for knee joint replacement.

## BIOMECHANICS OF LIGAMENTS AND TENDONS

Ligaments and tendons are composed primarily of collagen fibers, elastic fibers, and reticulum fibers. Ligaments and tendons derive their strength from collagen fibers and elasticity from elastic fibers. Tendons primarily experience unidirectional loads as muscles pull on them and as a result, the collagen fibers in tendons are aligned parallel to the direction of muscle loading. Ligaments receive loads in a variety of directions and as a result, the collagen fibers in ligaments are aligned in the primary directions of loading, with many fibers running in oblique angles. A typical stress–strain curve for a ligament or tendon is shown in Figure 15.21.

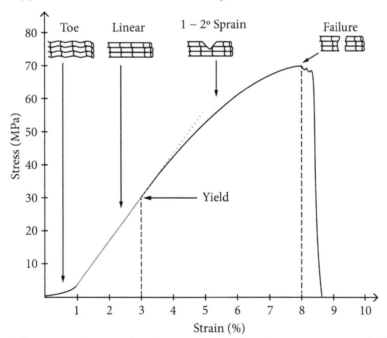

**Figure 15.21** Typical stress-strain curve for a ligament or tendon. Stress-strain curves for ligaments and tendons consist of four regions: toe, linear, yield, and failure. Adpated from Butler et al. 1978, Exer Sport Sci Rev, 6:145.

The ligament stress-strain curve in Figure 15.21 depicts the relationship between the ligament fiber bundles and the stress-strain curve properties. There are four distinct regions in the ligament stress-strain curve: toe, linear, yield, and failure. When a ligament is initially loaded, (toe region), the material is compliant and it readily deforms. The ligament fibers in an unloaded ligament have a wavy or crimp-like appearance when viewed under a microscope. The initial loading of a ligament tightens the ligament, causing the ligament crimp to be straightened out. Once the ligament fibers have been straightened out, the ligament stress-strain properties transition to the linear region of the curve. In the linear region of the stress-strain curve, the mechanical response of the ligament material is linear. The maximum stiffness of the ligament is determined by computing the slope of the stress-strain curve for the ligament in the linear region. The yield point for ligaments and tendons occurs at approximately 3% strain. Notice in Figure 15.21 that the slope of the stress-strain curve is less in the yield region than in the linear region of the curve, and therefore, ligaments are stiffer in the linear region. The yield region of the stress-strain curve includes strains above 3% up to the point of failure which can vary from 8 – 15%. Sports medicine professionals use the terms 1st, 2nd, and 3rd degree sprains to describe the grade of injury to a ligament. In a 1st degree sprain only a few ligament fibers are torn. For a 2nd degree sprain anywhere from a one-third to almost all of the ligament fibers are torn. In a 3rd degree sprain all of the fibers are torn. In terms of the stress-strain curve, second degree sprains are in the yield region of the curve and third degree sprains are in the failure region of the curve.

Since ligaments and tendons are viscoelastic materials, they will exhibit both time-dependent and velocity-dependent responses. Improving range of motion by participating in a stretching program is an example taking advantage of the time dependent response of ligaments and tendons. Holding a stretch for 30 seconds or more and repeatedly stretching will eventually fatigue ligament and tendon collagen cross-links causing them to break and reform at a longer length, thus improving range of motion. The velocity-dependent response of ligaments and tendons results in lower stiffness and strength when the material is loaded at a slow rate and higher stiffness and strength when loaded at a higher rate.

Ligaments and tendons are dynamic tissues that respond to the loads imposed upon them by adjusting the orientation of the collagen and elastic fibers to alter their mechanical properties. An example of this dynamic response can be seen in individuals who have had a surgical repair of a ligament. Physicians sometimes use either a portion of a hamstring tendon or patella tendon to repair a torn ACL ligament. Initially, the collagen fibers in the tendon that was used to repair the ligament primarily run along the long axis of the tendon, since tendons experience linear tension loading from muscles. Now that the former tendon is called upon to serve as an ACL ligament, it will experience loading in a wider variety of directions. As a result, over time, the collagen fibers in the tendon material that is now serving as an ACL ligament will reorient in oblique directions so that the material can now respond, all directions of loads imposed.

## EFFECTS OF EXERCISE ON CARTILAGE, LIGAMENTS, AND TENDONS

Wolff's Law can easily be extrapolated to ligaments, tendons, and cartilage. Appropriate levels of exercise will cause cartilage, ligaments and tendons to strengthen and essentially become thicker (Woo, et al., 1997). A lack of exercise will cause the reverse response, resulting in weaker ligaments, tendons, and cartilage. Figure 15.22 shows the effects of exercise on ligaments and tendons. In 1987, Savio Woo demonstrated that 9 weeks of immobilization of a healthy rabbit caused the MCL ligament to reduce ligament stiffness and strength to failure by one-third, when compared to a non-immobilized animal (Woo, et al., 1987). In addition,

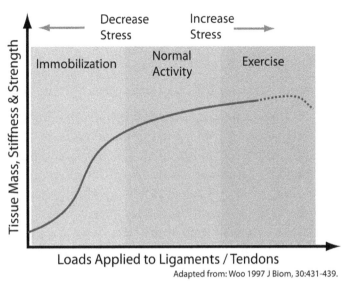

**Figure 15.22** Effects of exercise on the strength of ligaments and tendons.

he reported that it took 1 year of remobilization after the 9 weeks of immobilization for the ligament to recover normal strength and integrity. In a later study, he demonstrated the effects of exercise on increasing the strength of the MCL ligament, where he reported that a 12-month exercise program increased the strength to failure of the MCL ligament by 20% (Woo, et al., 1994).

The exercise effect on ligament healing and strength can be easily seen by looking at the historical change in the treatment of ACL injuries. In the 1970s and early 1980's, when an individual received an ACL injury that required surgery, the individual would be placed in a cast for up to 8 weeks. As outlined above, immobilization has adverse effects on the strength and stiffness of ligaments. Thus, when a football player tore and ACL in the early season it would take at least one year before the player was safely able to return to play. Today, ACL injuries are treated very differently. First, the individual is not placed in a cast, second, exercise is used even before the individual leaves the hospital to begin rehabilitating the injury. As a result of using exercise to rehabilitate injuries, athletes can now fully recover within 3–4 months.

## BIOMECHANICAL DESCRIPTORS OF THE MECHANISMS OF INJURY

The factors listed below describe the relative potential for injury and the magnitude of injury that might occur:

- Minimum time for neuromuscular system to respond to a stimulus
- Magnitude of force
- Rate of force application
- Point of force application

- Direction of force application
- Number of repetitions of loading
- State of training (de-trained, immobilized, over-trained)
- Critical limits of the tissue (warm-up, previous injuries)
- Magnitude of muscular preactivation

### Minimum Time for the Neuromuscular System to Respond to a Stimulus

When evaluating the relative potential for some unexpected event to cause an injury it is necessary to consider the minimum time for the neuromuscular system to respond to a stimulus, which can vary from 80–200 ms. The minimum time for the neuromuscular system to respond to a stimulus is comprised of two components: reflex time and muscle force time. Reflex time, which the latency between the application of the stimulus and the beginning of the muscles response (EMG), varies from 30–90 ms. Figure 15.23 gives an example of reflex response to an unexpected event. This device causes a sudden inversion about the ankle joint. As shown in Figure 15.23, when the platform falls and the ankle begins to invert, it takes the peroneus longus muscle 70 ms to begin to initiate a reflex response.

The next component of the minimum time for the neuromuscular system to respond to a stimulus is muscle force time. Depending upon the level of force required, it will take from 50–300 ms for the muscles involved to generate force.

Once the sensory system has initiated a reflex response and the muscle system has generated the required force to respond to the stimulus it may take additional time for the force to overcome the momentum associated with the movement. For example, imagine that you are playing basketball and you jump up to get a rebound. While you are up in the air, another player moves underneath you and as you come down the medial side of your foot lands on the edge of the other player's shoe, which then causes a rather large inversion torque. In order for your protective mechanisms to prevent an injury, the reflexes will need 30–90 ms and your muscles will need 300–500 ms to generate sufficient torque to resist the in-

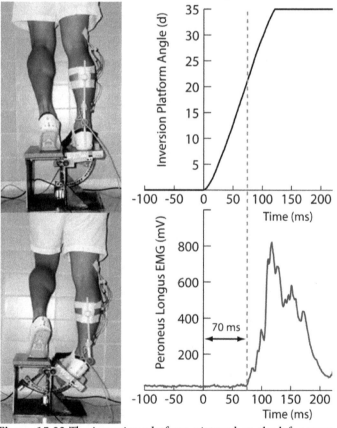

**Figure 15.23** The inversion platform pictured on the left causes the ankle to invert 35 d. EMG electrodes were placed on the peroneus longus muscle to measure the reflex latency.

version torque. Chances are it is too late; you probably sprained your ankle within the first 70 ms of landing on the opponent's shoe.

We often suggest that training will prevent injuries and that is true. A trained individual has stronger ligaments, tendons, muscles and cartilage, and as long as there is sufficient time to respond to some unexpected event, the trained individual will be more likely to be able to avoid an injury. As stated above, our estimate of the minimum time for the neuromuscular system to respond to a stimulus is 80–200 ms. Now to give an example that everyone has experienced. Imagine that it is dark and you are descending down a flight of stairs. You think you are at the bottom stair, but there is one more stair.

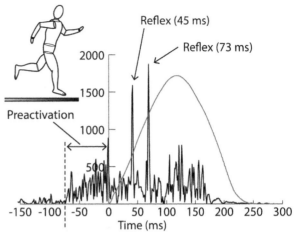

**Figure 15.24** Gastrocnemius (Ga) EMG and vertical force for barefoot running. Preactivation of Ga begins 70 ms prior to ground contact. Short and medium latency reflexes occur 45 and 73 ms after ground contact.

What happens? You free fall and crash onto the top of the last stair. You collide rather seriously with the stair and as soon as you realize that you are okay, you feel somewhat embarrassed. Additionally, you probably look around, hoping that no one saw you. The unexpected event occurred because you thought you were at the bottom and switched from the motor control program for stair descent to the motor control program for walking straight ahead on a level surface. If the neuromuscular system was really fast, it would have done something to save you.

## Muscular Preactivation and Expected Events

In the stair descent example above, the event involved was unexpected. The central nervous system (CNS) is optimized for efficiency in most movement situations. If you were told that when you went up to get a rebound, that someone's shoe would be underneath your foot as you were coming down, you would mobilize your body to prepare for the expected collision. This mobilization is referred to ask muscular preactivation. Since the neuromuscular system is relatively slow (80–200 ms), we preactivate the muscles of the joints involved before the known event will occur. The level of muscular preactivation is based upon previous experience. Usually, the muscles are activated about 100–200 ms before a known event. So when landing from a jump, the hip extensors, knee extensors, and plantar flexors are all activated about 100–200 ms before landing. Figure 15.24 shows muscular preactivation of the gastrocnemius and vertical ground reaction force for a barefoot runner. The gastrocnemius is preactivated 70 ms prior to ground contact. A short and medium reflex occurs 45 ms and 73 ms after ground contact.

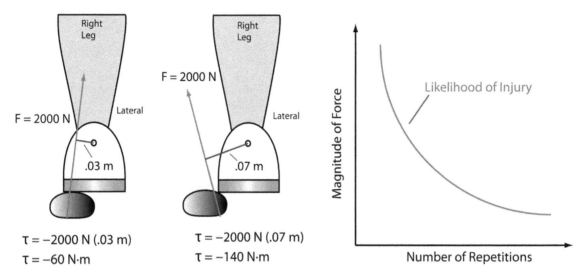

**Figure 15.25** Effects of point and direction of force on the torque about the joint center. On the left, the rock is positioned closer to the axis of rotation and the vector has a small moment arm, so it cause less torque than the position and direction for the rock on the right.

**Figure 15.26** A high magnitude of force may cause an injury in one repetition. When the magnitude of force is low, a large number of repetitions are needed to cause an injury.

## Magnitude of Force and Rate of Force Application

The magnitude of force has a significant effect on the potential for injury. If the magnitude of force exceeds the ability of the neuromuscular system to respond, an injury is inevitable. The rate of force application is directly rated to the minimum time of the neuromuscular system to respond. High magnitude forces that occur in less than 80 ms have a high potential to cause injury.

## Point and Direction of Force Application

Figure 15.25 gives an example of how altering the point and direction of force application can affect the likelihood of a force to cause an injury. In each case, a runner lands on a rock that is positioned underneath the medial side of the right leg. The force vector will cause a CW torque about the subtalar joint axis. In the drawing on the left, the point of application is closer to mid-line of the shoe and the direction of force application is 80 degrees. The resulting torque is –60 N·m. In the drawing on the right, the point of force application has been moved to the extreme medial edge of the shoe and the direction of force application is now 115 degrees. In this case the resulting torque is –140 N·m. This example illustrates how altering either the point of force application or the direction of force application can affect the likelihood of injury.

## Number of Repetitions

The relationship between the magnitude of force and the number of repetitions and the likelihood of a force to cause an injury is shown in Figure 15.26. On the left hand side of the curve, a high magnitude force may only require one repetition to cause an injury. On the right hand side of the curve, the magnitude of force is low and many repetitions may be required to cause an injury. Low magnitude force that cause an injury after a large number of repetitions are referred to as overuse injuries. Stress fractures are an example of an overuse injury. For example, in aerobic dance, the magnitude of force is low and the rate of force application is also low. Therefore, it is quite safe to assume that participating in aerobic dance once may not cause an injury. In contrast, imagine an aerobic dance teacher who teaches 5 classes per day, for 6 days per week. Eventually the teacher may get a stress fracture of the 5th metatarsal bone.

## Critical Limits of the Tissue and State of Training

The critical limits of the tissue include factors such as warming up. Failure to warm up before participating in a strenuous activity can often lead to an injury. Critical limits of the tissue are affected by the state of training, as a highly trained individual can withstand higher forces. Finally, the history of injury effects the critical limits of the tissue. A bone, ligament, or tendon that has a previous injury has irregularities within the tissue that affect the strength of the material. Ligaments and tendons fill in with scar tissue and scar tissue is stiffer than the uninjured tissue. In addition, repeated tears to ligaments and tendons can also cause the tendon or ligament to be elongated, which adversely affects its ability to protect the joint.

## REVIEW QUESTIONS

1. Describe the structural similarities and differences between cortical and cancellous bone.
2. What is a basic multicellular unit? Describe its role in remodeling.
3. What is muscular preactivation?
4. What is the minimum time for the neuromuscular system to respond to a stimulus? What factors affect the minimum time for the neuromuscular system to respond to a stimulus?
5. What methods were described to determine the quality and properties of bone?
6. Describe the role of cartilage in attenuating forces.
7. Describe the long term effect of impact forces on joints, as observed by Radin.

## REFERENCES

Allen M, Burr D. Skeletal microdamage: Less about biomechanics and more about remodeling. Clinical Reviews in Bone and Mineral Metabolism. 6: 24-30, 2008.

Balazs EA, Denlinger JL. Viscosupplementation: A new concept in the treatment of osteoarthritis. Journal of Rheumatology. 20: 3-9, 1993.

Bergström I, Landgren BM, Brinck J, Freyschuss B. Physical training preserves bone mineral density in postmenopausal women with forearm fractures and low bone mineral density. Osteoporosis International. 19: 177-183, 2008.

Biewener AA. Scaling body support in mammals: Limb posture and muscle mechanics. Science. 245: 45-48, 1989.

Brand R. Biographical sketch: Julius Wolff, 1836–1902. Clinical Orthopaedics and Related Research. 468: 1047-1049, 2010.

Burr DB, Robling AG, Turner CH. Effects of biomechanical stress on bones in animals. Bone. 30: 781-786, 2002.

Chen H, Zhou X, Shoumura S, Emura S, Bunai Y. Age- and gender-dependent changes in three-dimensional microstructure of cortical and trabecular bone at the human femoral neck. Osteoporosis International. 21: 627-636, 2010.

Chen J-H, Liu C, You L, Simmons CA. Boning up on Wolff's law: Mechanical regulation of the cells that make and maintain bone. Journal of Biomechanics. 43: 108-118, 2010.

Cowin SC. Wolff's law of trabecular architecture at remodeling equilibrium. Journal of Biomechanical Engineering. 108: 83-88, 1986.

Ehrlich PJ, Lanyon LE. Mechanical strain and bone cell function: A review. Osteoporosis International. 13: 688-700, 2002.

Forwood MR. Physical activity and bone development during childhood: Insights from animal models. Journal of Applied Physiology. 105: 334-341, 2008.

Fritton SP, J. McLeod K, Rubin CT. Quantifying the strain history of bone: Spatial uniformity and self-similarity of low-magnitude strains. Journal of Biomechanics. 33: 317-325, 2000.

Frost HM. Bone "mass" and the "mechanostat": A proposal. Anatomical Record. 219: 1-9, 1987.

Frost HM. Why do marathon runners have less bone than weight lifters? A vital-biomechanical view and explanation. Bone. 20: 183-189, 1997.

Fyhrie DP, Carter DR. Femoral head apparent density distribution predicted from bone stresses. Journal of Biomechanics. 23: 1-10, 1990.

Gardinier JD, Townend CW, Jen K-P, Wu Q, Duncan RL, Wang L. In situ permeability measurement of the mammalian lacunar-canalicular system. Bone. 46: 1075-1081, 2010.

Guilak F, Alexopoulos LG, Upton ML, Inchan Y, Jae Bong C, Li CAO, et al. The pericellular matrix as a transducer of biomechanical and biochemical signals in articular cartilage. Annals of the New York Academy of Sciences. 1068: 498-512, 2006.

Hazenberg JG, Taylor D, Lee TC. The role of osteocytes and bone microstructure in preventing osteoporotic fractures. Osteoporos Int. 18: 1-8, 2007.

Lakin BA, Grasso DJ, Shah SS, Stewart RC, Bansal PN, Freedman JD, et al. Cationic agent contrast-enhanced computed tomography imaging of cartilage correlates with the compressive modulus and coefficient of friction. Osteoarthritis and Cartilage. 21: 60-68, 2013.

Lanyon LE. Functional strain in bone tissue as an objective, and controlling stimulus for adaptive bone remodelling. Journal of Biomechanics. 20: 1083-1093, 1987.

Lanyon LE, Rubin CT. Static vs dynamic loads as an influence on bone remodelling. Journal of Biomechanics. 17: 897-905, 1984.

Liu XS, Bevill G, Keaveny TM, Sajda P, Guo XE. Micromechanical analyses of vertebral trabecular bone based on individual trabeculae segmentation of plates and rods. Journal of Biomechanics. 42: 249-256, 2009.

Liu XS, Sajda P, Saha PK, Wehrli FW, Bevill G, Keaveny TM, et al. Complete volumetric decomposition of individual trabecular plates and rods and its morphological correlations with anisotropic elastic moduli in human trabecular bone. Journal of Bone and Mineral Research. 23: 223-235, 2008.

Liu XS, Sajda P, Saha PK, Wehrli FW, Guo XE. Quantification of the roles of trabecular microarchitecture and trabecular type in determining the elastic modulus of human trabecular bone. Journal of Bone and Mineral Research. 21: 1608-1617, 2006.

Liu XS, Zhang XH, Guo XE. Contributions of trabecular rods of various orientations in determining the elastic properties of human vertebral trabecular bone. Bone. 45: 158-163, 2009.

Maroudas AI. Balance between swelling pressure and collagen tension in normal and degenerate cartilage. Nature. 260: 808-909, 1976.

McKay HA, MacLean L, Petit M, MacKelvie-O'Brien K, Janssen P, Beck T, et al. "Bounce at the bell": A novel program of short bouts of exercise improves proximal femur bone mass in early pubertal children. Br J Sports Med. 39: 521-526, 2005.

Mosekilde L. Age-related changes in bone mass, structure, and strength – effects of loading. Zeitschrift für Rheumatologie. 59: I1-I9, 2000.

Radin EL, Parker HG, Pugh JW, Steinberg RS, Paul IL, Rose RM. Response of joints to impact loading-iii. Journal of Biomechanics. 6: 51-57, 1973.

Rieppo J, Hyttinen MM, Halmesmaki E, Ruotsalainen H, Vasara A, Kiviranta I, et al. Changes in spatial collagen content and collagen network architecture in porcine articular cartilage during growth and maturation. Osteoarthritis and Cartilage. 17: 448-455, 2009.

Rochefort GY, Pallu S, Benhamou CL. Osteocyte: The unrecognized side of bone tissue. Osteoporosis International. 21: 1457-1469, 2010.

Rubin C, Gross TS, Qin Y-X, Fritton S, Guilak F, McLeod K. Differentiation of the bone-tissue remodeling response to axial and torsional loading in the turkey ulna. J Bone Joint Surg Am. 78: 1523-1533, 1996.

Rubin C, Recker R, Cullen D, Ryaby J, McCabe J, McLeod K. Prevention of postmenopausal bone loss by a low-magnitude, high-frequency mechanical stimuli: A clinical trial assessing compliance, efficacy, and safety. Journal of Bone and Mineral Research. 19: 343-351, 2004.

Rubin CT, McLeod KJ, Bain SD. Functional strains and cortical bone adaptation: Epigenetic assurance of skeletal integrity. Journal of Biomechanics. 23: 43-54, 1990.

Schurz J, Ribitsch V. Rheology of synovial fluid. Biorheology. 24: 385-399, 1987.

Shen Y, Zhang Z-M, Jiang S-D, Jiang L-S, Dai L-Y. Postmenopausal women with osteoarthritis and osteoporosis show different ultrastructural characteristics of trabecular bone of the femoral head. BMC Musculoskeletal Disorders. 10: 35, 2009.

Sissons HA, O'Connor P. Quantitative histology of osteocyte lacunae in normal human cortical bone. Calcif Tissue Res. 22 Suppl: 530-533, 1977.

Stauber M, Rapillard L, van Lenthe GH, Zysset P, Müller R. Importance of individual rods and plates in the assessment of bone quality and their contribution to bone stiffness. Journal of Bone and Mineral Research. 21: 586-595, 2006.

Vico L, Collet P, Guignandon A, Lafage-Proust MH, Thomas T, Rehaillia M, et al. Effects of long-term microgravity exposure on cancellous and cortical weight-bearing bones of cosmonauts. Lancet. 355: 1607-1611, 2000.

Weiss C, Rosenberg L, Helfet AJ. An ultrastructural study of normal young adult human articular cartilage. J Bone Joint Surg Am. 50: 663-674, 1968.

Wolff J. The classic: On the inner architecture of bones and its importance for bone growth. Clinical Orthopaedics and Related Research. 468: 1056-1065, 1870.

Woo SL, Chan SS, Yamaji T. Biomechanics of knee ligament healing, repair and reconstruction. Journal of Biomechanics. 30: 431-439, 1997.

Woo SL, Gomez MA, Sites TJ, Newton PO, Orlando CA, Akeson WH. The biomechanical and morphological changes in the medial collateral ligament of the rabbit after immobilization and remobilization. Journal of Bone and Joint Surgery. 69-A: 1200-1211, 1987.

Woo SL, Smith BA, Johnson GA. Biomechanics of knee ligaments. In: Fu FH, Harner CD, Vince KG, editors. Knee surgery. Baltimore: Williams & Wilkins, 1994.

Xia Y. Averaged and depth-dependent anisotropy of articular cartilage by microscopic imaging. Seminars in Arthritis and Rheumatism. 37: 317-327, 2008.

Zheng S, Xia Y. The collagen fibril structure in the superficial zone of articular cartilage by μmri. Osteoarthritis and Cartilage. 17: 1519-1528, 2009.

## SUGGESTED SUPPLEMENTAL SOURCES

Enoka, R. M. (2008). Neuromechanics of Human Movement. Champaign, Human Kinetics.

Franklin, K., P. Muir, et al. (2010). Introduction to Biological Physics for the Health and Life Sciences. Hoboken, John Wiley & Sons.

Griffiths, I. W. (2006). Principles of Biomechanics & Motion Analysis. Philadelphia, Lippincott Williams & Wilkins.

Hall, S. J. (2007). Basic Biomechanics. Boston, McGraw Hill.

Hamill, J. and K. M. Knutzen (2009). Biomechanical Basis of Human Movement. Philadelphia, Lippincott Williams & Wilkins.

McGinnis, P. M. (2013). Biomechanics of Sport and Exercise. Champaign, Human Kinetics.

Nordin, M. and V. H. Frankel (2001). Basic Biomechanics of the Skeletal System. Philadelphia, Lippincott Williams & Wilkins.

Whiting, W. C. and R. F. Zernicke (1998). Biomechanics of Musculoskeletal Injury. Champaign, Human Kinetics.

# Chapter 16

## MUSCLE EXCITATION-CONTRACTION COUPLING

### LEARNING OUTCOMES

**After reading Chapter 16, you should be able to:**

- Describe myofilament and sarcomere structure.
- Describe the role of cytoskeletal structures on muscle stiffness.
- Describe excitation-contraction coupling in skeletal muscle.
- Describe the historical development of the sliding filament theory.
- Describe the crossbridge cycle for concentric, eccentric, and isometric contractions.

### MUSCLE STRUCTURE

Three different layers of connective tissue effectively link muscle fibers together. The inner most layer is known as the endomysium and it surrounds individual muscle fibers. The second layer, the perimysium encapsulates bundles of muscle fibers into fascicles. Finally, the outermost layer, the epimysium surrounds the fascicles to encase the entire muscle. This connective tissue transmits the tension developed within the individual muscle fibers to the tendon.

The basic structure of muscle fiber is shown in the Figure 16.1. A skeletal muscle fiber is an elongated, cylindrical cell with a diameter of 10–60 μm and a length 10–300 mm. A single muscle fiber contains hundreds of myofilaments, which are bundled together and surrounded by an excitable cell membrane called the sarcolemma. The neuromuscular junction is located just above the sarcolemma. The neuromuscular junction includes the gap from the alpha motor neuron's endplate, or terminal branches to the muscle membrane. The transverse tubule, also called the T-tubule is directly underneath the cell membrane. The transverse tubule system transmits the muscle fiber activation message from the cell membrane into the muscle fiber. The T-tubule runs perpendicular thru the muscle fiber, where it makes connections with the sarcoplasmic reticulum. The sarcoplasmic reticulum stores calcium ions, which are used in the regulation of muscle contraction.

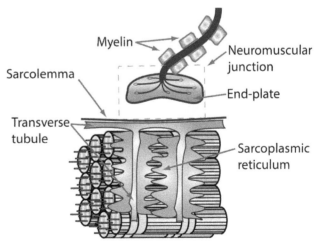

**Figure 16.1** Structure of a single muscle fiber and the neuromuscular junction.

## Cytoskeleton Structure

Inside the muscle cell there is a microscopic network of protein filaments that are collectively referred to as the cytoskeleton. Two of these cytoskeleton structures, actin and myosin, are well known. Actin and myosin are the contractile components of the muscle machine. As we shall soon see, a subfragment of the myosin arm (S1) is the actual molecular motor of muscular force generation. The force generated by the molecular motors requires a cellular framework to transmit the tension to the tendon. The cytoskeleton contains three molecular springs that serve as the necessary framework of force transmission: costamere, desmin, and titin (Carlsson, et al., 2008; Patel, et al., 1997). Figure 16.2 depicts the cytoskeleton structure of muscle. As shown in Figure 16.2, costamere is an elastic spring that connects the z disk (z line) to the muscle membrane (sarcolemma). Desmin connects adjacent filament bundles to the z disk. Finally, titin connects the myosin filament to the z disk. In terms of force transmission, desmin and costamere transmit force perpendicular to the muscle fiber, ensuring that the entire muscle fiber will shorten or lengthen approximately in unison. Titin, on the other hand, transmits the force longitudinally down the muscle fiber. These molecular springs (titin, desmin, and costamere) contribute to the passive stiffness of muscle and to the braking action of muscle during eccentric contractions.

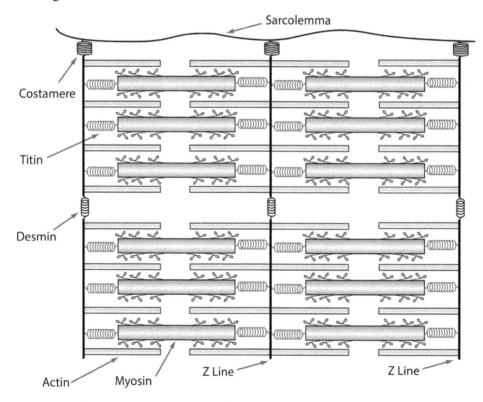

**Figure 16.2** Structure of muscle cytoskeleton. In addition to the contractile proteins (actin and myosin) the muscle cytoskeleton consists of three molecular springs (costamere, desmin, and titin). The costamere elastic element connects the z line (z disk) to the sarcolemma. Desmin connects adjacent fiber bundles at the z line and titi connects myosin to the z line.

## MYOFILAMENT STRUCTURE

The myofilaments are primarily composed of the two protein structures myosin and actin, shown in Figure 16.3. Myosin is also referred to as the thick filament and actin is also known as the thin filament. The thin filament is composed of two helical strands of actin, a slender rod tropomyosin, a globular structure called troponin, and an actin binding site capable of binding with myosin. Troponin and tropomyosin are both regulatory proteins. Under resting conditions, they prevent a crossbridge from forming between actin and myosin. Troponin can reversibly bind with $Ca^{++}$ ions and the amount of binding is directly related to the calcium concentration.

Myosin, the thick filament, is composed of a long, slender rod-like structure with a large number of myosin arms extending outward three-dimensionally. The myosin arm is composed of two subfragments: S1 and S2 (shown in the lower left of Figure 16.3). The S1 subfragment has two globular heads which contain ATP. During actomyosin crossbridge formation, one or both of the myosin heads attach to actin at the actin binding site. The **S1 subfragment**, also referred to as the "**molecular motors**", contains the chemical and mechanical machinery needed for the myosin arm to generate tension in muscular contraction. During the power stroke phase of the actomyosin crossbridge cycle, the S1 subfragment rotates 70 degrees around the neck region of the myosin arm, causing the fiber to shorten approximately 10 nm (Shih, et al., 2000).

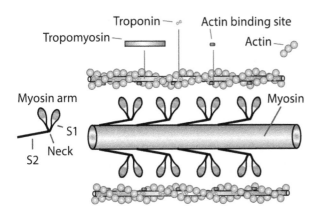

**Figure 16.3** Structure of the muscle proteins actin and myosin. The actin filament contains troponin, tropomyosin and an actin binding site. The myosin filament contains a repeating series of myosin arms. Each myosin arm is composed of an S1 and S2 subfragment. The globular heads on S1 are the molecular motors that rotate about the neck region during muscular contraction.

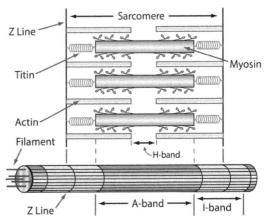

**Figure 16.4** The sarcomere contains the proteins actin, myosin, and titin. Titin is an elastic protein that connects myosin to the z disk. The H-band contains only myosin. The A-band contains actin and myosin. The I-bands contain primarily actin.

## Sarcomere – The Functional Unit

When viewed under a microscope, skeletal muscle consists of a series of light and dark repeating sections, called the sarcomere. The sarcomere is often said to be the functional unit of the muscle cell. Muscular tension is developed in each sarcomere and the tension is transmitted longitudinally down the muscle fiber, as well as perpendicular to the direction of the muscle fiber by the connective tissue described earlier. The sarcomere, shown in Figure 16.4, is defined from Z line to Z line and it includes myosin, actin, and titin. Titin is an elastic protein structure that connects myosin to the Z lines, also called the Z disks. Titin transfers the force generated in the actomyosin crossbridge longitudinally down the muscle fiber where it eventually reaches the tendon. The A-band, shown in Figure 16.4, contains both actin and myosin filaments. The H-band contains only the thick (myosin) filaments. When the sarcomere shortens and the filaments overlap, the H-band disappears. The lighter regions between the darker A-bands are called I-bands and they contain primarily actin.

## Excitation–Contraction Coupling

Muscular contraction is initiated at the spinal cord level, where the neural input to the alpha motor neuron causes a stimulus to be propagated down the nerve fiber (see Figure

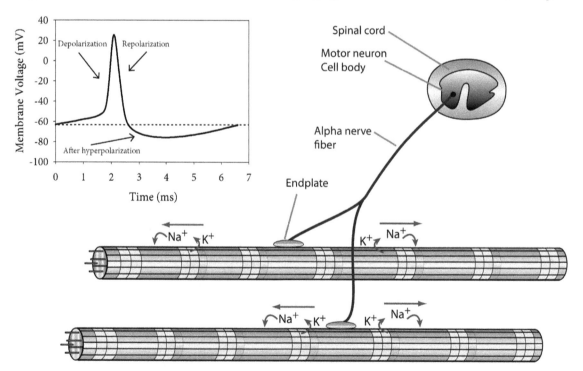

**Figure 16.5** Muscle fiber action potential. The muscle fiber action potential is propagated in each direction along the muscle fiber. Acetylcholine crosses the synapse causing $Na^+$ And $K^+$ ion channels to open in the muscle membrane. $Na^+$ flows into the cell causing the muscle membrane potential to go from −60 to +30 mV, see graph of muscle fiber action potential (upper left). $K^+$ is then pumped out of the cell to return the membrane back to negative resting potential.

16.5). When the stimulus arrives at the endplate, the neurotransmitter acetylcholine (ACh) is released from the endplate. Once ACh crosses the synapse, it causes $Na^+$ and $K^+$ channels in the muscle membrane to open up and a muscle fiber action potential is generated in each direction along the muscle fiber, depicted by the arrows in Figure 16.5. In response to the opening of the $Na^+$ and $K^+$ ion channels in the muscle membrane opening up, $Na^+$ flows into the cell and $K^+$ flows out of the cell. The normal resting potential for the muscle membrane is $-60$ mV (see membrane voltage graph in the upper left of Figure 16.5). The concentration gradient for $Na^+$ is less than the concentration gradient of $K^+$ ions and as a result $Na^+$ flows into the cell at a faster rate than $K^+$ exits. This causes the membrane voltage to attain a positive value. Within a few milliseconds, the muscle membrane is returned to its negative resting value by closing the $Na^+$ channels and pumping $K^+$ to the cell exterior.

Excitation–contraction coupling is the term used to describe the sequence of events that occur after a muscle fiber action potential is generated until the muscle fiber force generates tension. In Figure 16.6, the muscle fiber action potential ($Na^+$ in, $K^+$ out) is propagated along the muscle surface and then it spreads downward into the transverse tubule system. The arrival of the muscle fiber action potential down into the T-tubule causes $Ca^{++}$ to be released from the sarcoplasmic reticulum. Under resting conditions, the two regulatory proteins troponin and tropomyosin inhibit myosin from forming a crossbridge with actin. Once $Ca^{++}$ is released from the sarcoplasmic reticulum, it binds with troponin. The binding of $Ca^{++}$ with troponin causes a change in the configuration of the actin filament. In the resting state, tropomyosin covers the actin binding site and prevents (inhibits) the formation of a crossbridge. The change in configuration of the actin filament exposes the actin binding site and since myosin has a high affinity to bond with actin, a crossbridge between actin and myosin is formed. The configurational change of actin removes the normal inhibition of actin–myosin bonding that is regulated by troponin and tropomyosin and is referred to as **disinhibition**. The myosin motor (S1 globular head) contains ATP, which is downgraded to ADP plus a high-energy phosphate (Pi). From the chemical energy obtained in this reaction, the motor rotates and generates tension. The next section will examine the mechanical and chemical events in an actomyosin crossbridge in greater detail.

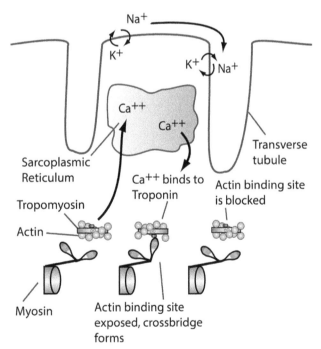

**Figure 16.6** Excitation–contraction coupling. $Ca^{++}$ binds to troponin, causing tropomyosin to move away from the actin binding site, enabling the formation of an actomyosin crossbridge.

## MECHANICAL AND CHEMICAL EVENTS IN THE CROSSBRIDGE CYCLE

The mechanical and chemical events in an actomyosin crossbridge cycle are shown in Figure 16.7. Figure 16.7 relates the chemical states of the myosin motor to the mechanical events that are triggered in each step of the crossbridge cycle. The myosin motor utilizes the energy obtained from the hydrolysis of ATP to do mechanical work on the actin filament. The chemical events in the cycle determine when an actomyosin crossbridge is formed and the state of binding (unbound, weakly-bound, strongly-bound) of the myosin motor to actin. The cycle begins at (a), with an unattached myosin motor containing ATP. As shown in Figure 16.7a, the myosin motor is "reprimed" and rotated back from after previous cycle, which is known as the pre-power stroke phase. In the next step (b), the hydrolysis of ATP begins (ADP•Pi) and the motor forms a weakly-bound crossbridge with actin. In step (c), the hydrolysis of ATP to ADP + Pi is complete and now the motor is strongly bound to actin. The power stroke (d), is

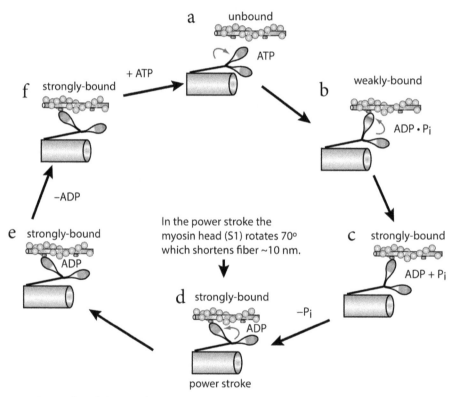

**Figure 16.7** Mechanical and chemical events in an actomyosin crossbridge cycle. (a) The cycle begins with unattached myosin containing ATP and the myosin arm is "reprimed", rotated back from after previous cycle. (b) Once the hydrolysis of ATP begins (ADP•Pi), myosin forms a weakly-bound crossbridge with actin. (c) When the hydrolysis of ATP to ADP + Pi is complete, myosin is strongly-bound to actin. (d) The power stroke begins when the Pi is released, causing the myosin neck (S1) to rotate 70°, shortening the fiber ~ 10 nm. (e) At the end of power stroke the crossbridge is still tightly bound. (f) When the ADP is released myosin is still strongly-bound to actin in rigor state. (f) When myosin picks up a new ATP, it reprimes the myosin S1 head by rotating it back to the pre-power stroke position and the next cycle begins at (a).

Adapted from Geeves et al. 2005 Cell Mol Life Sci, 62:1462-1477.

initiated when the inorganic phosphate (Pi) is released, causing the motor (S1) to rotate 70° about the neck of the myosin arm (Shih, et al., 2000). This rotation of the motor shortens the muscle fiber by approximately 10 nm as the actin filament is pulled toward the center of the sarcomere. The motor converts the chemical energy obtained from ATP into mechanical energy that is used to do work on actin. In step (e), the power stroke has concluded and the motor is still strongly-bound to actin. Following (e) the ADP is released and in step (f), the motor is still strongly bound to actin. Step (f) is referred to as the rigor state. Finally, the rigor state is terminated when the myosin motor picks up a new ATP, which reprimes the motor, causing S1 to rotate back to the pre-power stroke position, where the next cycle begins at (a).

## HISTORICAL DEVELOPMENT OF ACTIN–MYOSIN CROSSBRIDGE THEORY

The crossbridge theory of muscular contraction is attributed to two scientists with the same last name, Hugh Huxley and Andrew Huxley. Hugh Huxley was the first to suggest that actin and myosin filaments move relative to each other during muscular contraction, the so-called "sliding filament theory" (Huxley, 1953). Based upon the earlier work of Hugh Huxley (Huxley, et al., 1954) and Andrew Huxley (Huxley, et al., 1954), the sliding filament theory was originally proposed by Andrew Huxley in 1957 (Huxley, 1957). We now believe that during the working or power stroke, the myosin S1 motor rotates thru an angle of approximately 70 degrees, which causes the fiber to shorten by 10 nm (Shih, 2000) and that each motor is capable of producing a maximum force of 6 pN (Piazzesi, et al., 2007).

### CONCENTRIC ACTOMYOSIN CROSS-BRIDGE CYCLE

The concentric actomyosin crossbridge cycle is shown in Figure 16.8. The letters A–D shown on the left of the figure give the time sequence for the concentric actomyosin crossbridge cycle. The cycle begins in A with the motor detached in the premotor state. In step B the motor is weakly attached and then following the dissociation of Pi the motor becomes strongly attached. Between steps B and C, the inorganic phosphate Pi is released and the power stroke occurs causing the motor to rotate 70° and

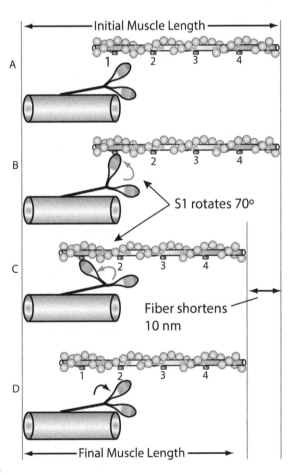

**Figure 16.8** Concentric actomyosin crossbridge cycle. The motor (S1) rotates 70°, causing the fiber to shorten 10 nm. In a concentric contraction, the motor generates 4-6 pN of force.

shortening the fiber by 10 nm. Finally, in step D, the motor picks up a new ATP and the motor rotates back to the premotor state, where the next crossbridge will occur at site 2 on the actin filament. The force produced by the molecular motor varies linearly with the velocity of shortening. At very slow velocity, a motor produces 6 pN of force. In very high velocity shortening, the motor force declines to 4 pN (Piazzesi, et al., 2007). In summary, in a single concentric actomyosin crossbridge cycle, the molecular motors yield 4–6 pN of force at a metabolic cost of one ATP.

Some authors have suggested that the second myosin motor attaches to the next available actin binding site when the velocity of shortening is at intermediate values (Brunello, et al., 2007; Mansson, 2010). It is widely understood that muscles generate the greatest concentric power at intermediate velocities. Mansson (2010) demonstrated that a rate-dependent attachment of the second motor in mathematical modeling yielded the best fit with experimental data. In addition, he suggested that, due to the low power production observed experimentally in very slow and very fast concentric contractions, attachment of the second myosin motor would simply be inefficient, using ATP with little or no additional force production. While this second motor attachment theory is possible, it is equally likely that increased muscle fiber stiffness and elastic contributions from cytoskeletal springs (costamere, desmin, and titin) may explain the enhanced power production at intermediate velocities. Evidence of the role of crossbridge and elastic element stiffness is presented in the next section.

## NON-LINEAR MUSCLE FIBER STIFFNESS

The muscle cytoskeletal structure is composed of elastic structures including the contractile machinery (actin, myosin) and the force-transmitting connective tissues (costamere, desmin, and titin), see Figure 16.2. These biological springs have been shown to be non-Hookean (Edman, 2009). The graph shown on the right of Figure 16.9 illustrates the deviation from linearity of muscle fiber stiffness as a function of the percentage of isometric force. The dashed line in the graph illustrates the expected linear response of an ideal spring. According to Hooke's law, there is a linear relationship between the stiffness of the spring and the force required to deform the spring. Edman found that muscle fiber stiffness increases exponentially with increasing isometric force.

The elastic components that contribute to the observed non-linear changes in muscle fiber stiffness are typically illustrated using a spring on the S2 subfragment of the myosin arm, as shown on the left in Figure 16.9. The actual anatomical location of the non-linear stiffness is thought to be primarily attributed to elastic deformation of: the myosin motor, titin, desmin, and costamere. In addition, the attachment of the second motor in the weakly-bound state has also been suggested to contribute to observed non-linear stiffness. The actomyosin crossbridge shown on the left in Figure 16.9 illustrates how elastic strain is thought to cause non-linear stiffness. In the top image, a myosin motor is generating a power stroke and attempting to rotate and shorten the muscle fiber. During this power stroke, an external force is resisting the shortening of the muscle fiber, which strains the elastic elements and also causes the second motor to attach in a weakly-bound state. The combination of elastic strain and the attachment of the second motor increases stiffness.

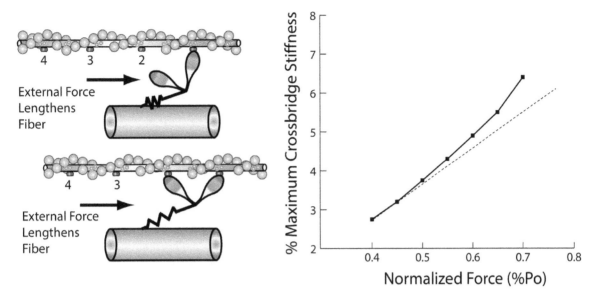

**Figure 16.9** Non-linear stiffness of crossbridges and elastic cytoskeletal structures. The cytoskeletal structures (titin, desmin, costamere, and the myosin motor) produce an exponential increase in muscle fiber stiffness with an increase of muscle force. Crossbridge stiffness data adapted from Edman 2009 J Exp Biol, 212: 1115-1119.

## ISOMETRIC ACTOMYOSIN CROSSBRIDGE CYCLE

The isometric actomyosin crossbridge cycle is shown in Figure 16.10. The letters A–E shown on the left of the figure give the time sequence for the isometric actomyosin crossbridge cycle. The cycle begins in A with the myosin motor detached in the premotor state. In step B, the motor is weakly attached and then following the dissociation of Pi, the motor becomes strongly attached by the end of step B. Between steps B and C, the inorganic phosphate Pi is released and the power stroke is initiated. However, in an isometric contraction, an external force or object resists changes in length of the muscle fiber. Therefore, initially, the motor is strained, as shown in step C. From the strained position in C, the motor rotates 70° and shortening the fiber by 6–10 nm. From steps B to C, the fiber is lengthened, and from the lengthened position in step C, the motor rotates and shortens the fiber. During the power stroke (steps C and D), the individual actomyosin motor generates 6 pN of force (Colombini, et al., 2007). However, due to increased fiber stiffness from elastic strain, 10 pN of force is produced. The amount of fiber shortening from steps A to D is approximately 0 nm, because the fiber is lengthened from steps B-C and shortened from steps C-D. In step E, the myosin motor picks up a new ATP, rotates backward and the isometric actomyosin cycle ends. In summary, the metabolic cost for this single isometric actomyosin crossbridge cycle was one ATP, it generated 10 pN of force per actomyosin bond and the muscle fiber length remains constant. The net force of 10 pN in a single isometric actomyosin crossbridge cycle arises from the active motor force of 6 pN and an additional force of 4 pN from elastic strain of the molecular springs (myosin arms, actin, titin, desmin, and costamere).

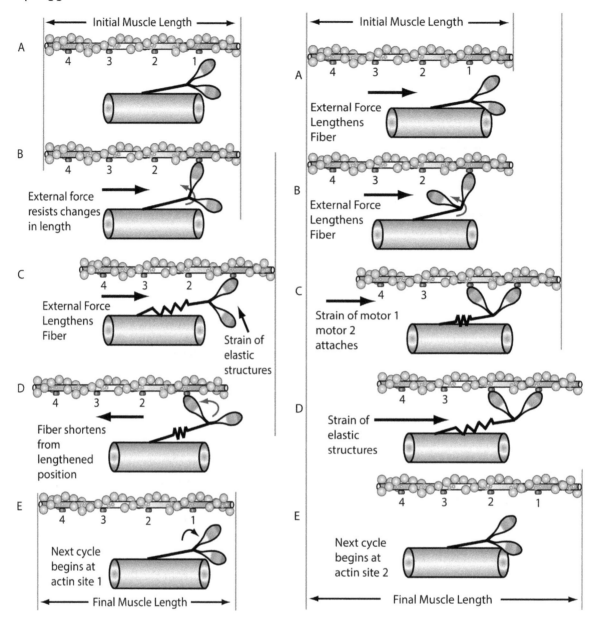

**Figure 16.10** Isometric actomyosin crossbridge cycle. The crossbridge generates 10 pN of force and uses 1 ATP in an isometric crossbridge cycle. The net fiber length is constant. The myosin motor generates 6 pN of force and the passive strain of 4 pN comes from enhanced stiffness of cytoskeleton.

**Figure 16.11** Eccenctric actomyosin crossbridge cycle. The crossbridge generates 10 pN of force and uses 1 ATP in an isometric crossbridge cycle. The myosin motor generates 6 pN of force and a passive strain of 4-14 pN comes from enhanced stiffness of cytoskeleton. Fiber length increases by 10 nm.

## ECCENTRIC ACTOMYOSIN CROSSBRIDGE CYCLE

In an eccentric contraction, an external force causes the muscle fiber to lengthen under tension. Landing from a vertical jump is an example of an external force that will cause the muscle fibers to lengthen under tension. The myosin motors attempt to shorten the muscle fiber, but the external force is greater than the force generated by the motors, which then strains both the crossbridges and elastic cytoskeletal structures. The crossbridges and cytoskeletal structures act as a brake to resist lengthening of the muscle. This forced lengthening of active motors and strain of elastic structures enables the motors to resist with forces that are two times greater than maximum isometric force.

The eccentric actomyosin crossbridge cycle is shown in Figure 16.11. The letters A–E shown on the right of the figure give the time sequence for the eccentric crossbridge cycle. The cycle begins in A with the myosin motor detached in the premotor state. In step B the motor is weakly attached and then following the dissociation of Pi, the motor becomes strongly attached by the end of step B. Between steps B and C, the inorganic phosphate Pi is released and the power stroke is initiated. In the power stroke, the motor attempts to rotate to shorten the muscle fiber by generating a motor force of 6 pN. Since the external force exceeds the motor force of 6 pN, the fiber undergoes forced lengthening. In step C, the muscle fiber stiffness is increased when the second motor attaches. The external force continues the fiber elongation, causing further strain of the: myosin motor, titin, desmin, and costamere elastic structures (step D). In addition to the active motor force of 6 pN, the molecular springs (myosin arms, actin, titin, desmin, and costamere) resist lengthening, which increases muscle stiffness (Nocella, et al., 2013). As a result of both the active motor force and the elastic strain, the eccentric power stroke generates 10–20 pN of force. Finally in step E, the motor picks up a new ATP and rotates back to the pre-motor state. The next crossbridge will form at site 2 on the actin filament.

The myosin motor and molecular springs resist lengthening with a force of 10–20 pN over an average distance of 1.25 nm (Colombini, et al., 2010). This resistive or braking force increases with increasing velocity of stretch. In slow velocity lengthening (velocity less than 0.5 μm/s per half sarcomere) the eccentric force increases rapidly as a function of increasing the velocity of stretch (Getz, et al., 1998). The force is constant at approximately 2 times the maximum isometric force (Po) for eccentric velocity above 0.5 μm/s per half sarcomere (Getz, et al., 1998). Most of this additional force is due to elastic strain, with only about 11% due to the attachment of the second myosin motor (Nocella, et al., 2013). Eccentric contractions are very efficient. The eccentric actomyosin crossbridge cycle resists lengthening with a force of 10–20 pN at a metabolic cost of only 1 ATP.

## EFFICIENCY OF ECCENTRIC CONTRACTIONS

The strain of active myosin motors and elastic cytoskeletal structures enables the crossbridges to resist lengthening with a force of 1.2–2.0 times maximum isometric force (Po). Eccentric contractions are unique in that they generate very high muscular force, at a lower metabolic cost than either isometric, or concentric contractions (Abbott, et al., 1952; Perrey, et al., 2001; Bickham, et al., 2011). Bickham and co-workers demonstrated the efficiency of eccentric contractions by computing the rate of release of Pi during eccentric contractions, which is a direct measure of energy utilization (Bickham, et al., 2011). They found a sharp decrease in the rate of Pi release at the beginning of the eccentric stretch. The rate of Pi release in the eccentric contraction was one-third of the rate of Pi in an isometric contraction. This decreased rate of energy utilization was present throughout the entire eccentric phase. While they energy utilization was one-third of isometric value, the force exhibited a sharp increase for 10-20 ms at the beginning of the stretch followed by a slower rate of force increase as the eccentric stretch continued.

In summary, eccentric contractions are much more efficient than isometric, or concentric contractions. Eccentric contractions produce higher force (1.2–2.0 times greater than isometric) while using only one-third of the energy required for an isometric contraction.

## SUMMARY OF ACTOMYOSIN CROSSBRIDGE CYCLE

Table 16.1 summarizes the actin-myosin crossbridge cycle for concentric, eccentric and isometric contractions.

**Table 16.1** Summary of Actomyosin Crossbridge Cycle

| Contraction Type | Energy Utilized | Force Generated | Change in Fiber Length | Effect of Velocity on Force |
|---|---|---|---|---|
| Concentric | 1 ATP | 4–6 pN | Shortened ~10 nm | Force decreases with increasing velocity: 4 pN for fast velocity 6 pN for slow velocity |
| Eccentric | 1 ATP | 10–20 pN 6 pN from motors, 4–14 pN resistive force from elastic strain. | Lengthened ~10 nm | Force increases with increasing velocity: 10 pN for slow velocity 20 pN for faster velocity |
| Isometric | 1 ATP | 10 pN 6 pN from motors, 4 pN resistive force from elastic strain. | Relatively Constant | none |

## REVIEW QUESTIONS

1. What causes disinhibition in the actin and myosin to enable the formation of a crossbridge?
2. What structures are responsible for increased muscle stiffness with increasing force?

3. Who are the two scientists who were instrumental in the development of the sliding filament theory?

4. What is the range of force generated in a single concentric actomyosin crossbridge from the slowest to the fastest velocity of shortening?

5. How does the velocity of lengthening affect the force generated in a single eccentric actomyosin crossbridge cycle?

6. List the steps from the stimulation of a nerve to the production of tension.

7. What structure on the actin filament does the myosin head attach to in order to form a crossbridge?

8. What event causes the configurational change in the actin filament that exposes the actin binding site?

9. What makes the muscle fiber lengthen in an eccentric contraction?

## References

Abbott BC, Bigland B, Ritchie JM. The physiological cost of negative work. J Physiol. 117: 380-390, 1952.

Bickham DC, West TG, Webb MR, Woledge RC, Curtin NA, Ferenczi MA. Millisecond-scale biochemical response to change in strain. Biophysical Journal. 101: 2445-2454, 2011.

Brunello E, Reconditi M, Elangovan R, Linari M, Sun Y-B, Narayanan T, et al. Skeletal muscle resists stretch by rapid binding of the second motor domain of myosin to actin. Proceedings of the National Academy of Sciences. 104: 20114-20119, 2007.

Carlsson L, Yu J-G, Thornell L-E. New aspects of obscurin in human striated muscles. Histochemistry and Cell Biology. 130: 91-103, 2008.

Colombini B, Bagni MA, Romano G, Cecchi G. Characterization of actomyosin bond properties in intact skeletal muscle by force spectroscopy. Proc Natl Acad Sci U S A. 104: 9284-9289, 2007.

Colombini B, Nocella M, Benelli G, Cecchi G, Bagni MA. Cross-bridge properties in single intact frog fibers studied by fast stretches. Advances in Experimental Medicine and Biology. 682: 191-205, 2010.

Edman KAP. Non-linear myofilament elasticity in frog intact muscle fibres. J Exp Biol. 212: 1115-1119, 2009.

Geeves MA, Fedorov R, Manstein DJ. Molecular mechanism of actomyosin-based motility. Cellular and Molecular Life Sciences CMLS. 62: 1462-1477, 2005.

Getz EB, Cooke R, Lehman SL. Phase transition in force during ramp stretches of skeletal muscle. Biophysical Journal. 75: 2971-2983, 1998.

Huxley AF. Muscle structure and theories of contraction. Prog Biophys Biophys Chem. 7: 255-318, 1957.

Huxley AF, Niedergerke R. Structural changes in muscle during contraction. Nature. 173: 974-973, 1954.

Huxley HE. Electron microscope studies of the organisation of the filaments in striated muscle. Biochim Biophys Acta. 12: 387-394, 1953.

Huxley HE, Hanson JM. Changes in the cross-striations of muscle during contraction and stretch and their structural interpretation. Nature. 173: 973-976, 1954.

Lombardi V, Piazzesi G. The contractile response during steady lengthening of stimulated frog muscle fibres. J Physiol. 431: 141-171, 1990.

Mansson A. Actomyosin-adp states, interhead cooperativity, and the force-velocity relation of skeletal muscle. Biophysical Journal. 98: 1237-1246, 2010.

Nocella M, Bagni M, Cecchi G, Colombini B. Mechanism of force enhancement during stretching of skeletal muscle fibres investigated by high time-resolved stiffness measurements. Journal of Muscle Research and Cell Motility. 34: 71-81, 2013.

Patel TJ, Lieber RL. Force transmission in skeletal muscle: From actomyosin to external tendons. In: Holloszy JO, editor. Exercise and sport sciences reviews. Baltimore: Williams & Wilkins. p. 321-363, 1997.

Perrey S, Betik A, Candau R, Rouillon JD, Hughson RL. Comparison of oxygen uptake kinetics during concentric and eccentric exercise. Journal of Applied Physiology. 91: 2135-2142, 2001.

Piazzesi G, Francini F, Linari M, Lombardi V. Tension transients during steady lengthening of tetanized muscle fibres of the frog. J Physiol. 445: 659-711, 1992.

Piazzesi G, Reconditi M, Linari M, Lucii L, Bianco P, Brunello E, et al. Skeletal muscle performance determined by modulation of number of myosin motors rather than motor force or stroke size. Cell. 131: 784-795, 2007.

Shih WM, Gryczynski Z, Lakowicz JR, Spudich JA. A fret-based sensor reveals large atp hydrolysis–induced conformational changes and three distinct states of the molecular motor myosin. Cell. 102: 683-694, 2000.

## SUGGESTED SUPPLEMENTAL SOURCES

Enoka, R. M. (2008). Neuromechanics of Human Movement. Champaign, Human Kinetics.

Hall, S. J. (2007). Basic Biomechanics. Boston, McGraw Hill.

Hamill, J. and K. M. Knutzen (2009). Biomechanical Basis of Human Movement. Philadelphia, Lippincott Williams & Wilkins.

Kandel, E. R., J. H. Schwartz, et al. (2013). Principles of Neural Science. New York, McGraw-Hill.

Lieber, R. L. (1992). Skeletal Muscle Structure and Function. Implications for Rehabilitation and Sports Medicine. Baltimore, Williams & Wilkins.

McGinnis, P., M. (2013). Biomechanics of Sport and Exercise. Champaign, Human Kinetics.

Pollack G. H. (1990). Muscles & Molecules. Uncovering the Principles of Biological Motion. Seattle, Ebner & Sons Publishers.

Winter, D. A. (2009). Biomechanics and Motor Control of Human Movement. Hoboken, John Wiley & Sons, Inc.

# Chapter 17

## MUSCLE MECHANICS

### LEARNING OUTCOMES

After reading Chapter 17, you should be able to:
- Define the muscle force–length relationship.
- Define the muscle force–velocity relationship.
- Define the stretch–shorten cycle.
- Define the three component muscle model.

## METHODS USED TO DETERMINE MUSCLE FORCE–LENGTH AND

## FORCE–VELOCITY RELATIONSHIP

The experimental apparatus used to study the muscle force-length and force-velocity relationship is shown in Figure 17.1. The diagram on the top displays the arrangement used for a whole muscle and the diagram on the bottom shows a similar arrangement that is used for a muscle fiber. The muscle tissue (whole muscle or fiber) is mounted between a force transducer and a length transducer in a thermostatically controlled glass chamber that is filled with Ringer's solution. A servomotor used to control the rate of muscle lengthening or shortening. Sarcomere lengths during the experiment are determined using a laser diffraction technique, where a laser beam is passed through the muscle fiber and the resulting diffraction pattern is measured to determine sarcomere length (Flitney, et al., 1978; Hill, 1953; Huxley,

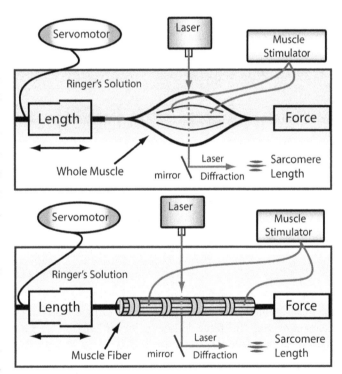

**Figure 17.1** Instrumentation used to study muscle mechanics for whole muscle (top), and single muscle fiber (bottom).

1957). The muscle tissue can be activated using a muscle stimulator. Once the muscle is activated the servomotor is used to lengthen or shorten the muscle at a fixed velocity while simultaneously measuring muscle length, muscle force, and sarcomere length.

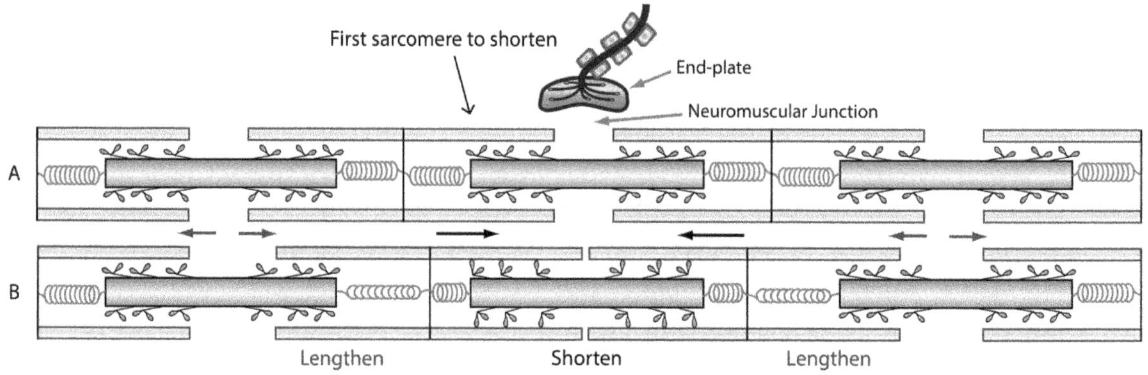

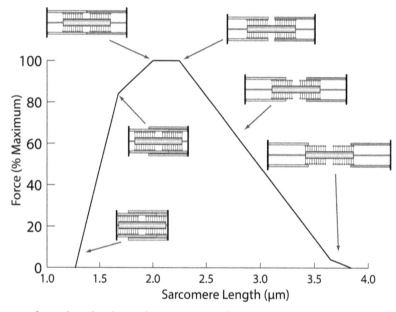

**Figure 17.2** Length changes to individual sarcomeres at the onset of an isometric contraction. A. Relaxed sarcomere. B. At the onset of muscular activation the center sarcomere which is directly underneath the endplate is the first sarcomere to shorten. When the center sarcomere shortens it pulls on the sarcomeres to the right and left causing them to lengthen.

**Figure 17.3** Sarcomere force–length relationship. Maximum force occurs at optimal length (2.0 – 2.2 μm) where the greatest number of crossbridges can be formed. The force declines as shorter fiber lengths (< 2.0 μm) due to filament overlap. The force declines at longer lengths (> 2.5 μm) because the myosin motors are unable to reach actin binding sites to form a crossbridge.

Adapted from Gordon AM, Huxley AF, Julian FJ. The variation in isometric tension with sarcomere length in vertebrate muscle fibres. Journal of Physiology. 184: 170-192, 1966. Used with permission from John Wiley & Sons.

The term isometric has generally been thought of as referring to a constant muscle length. However, due to muscle compliance, the muscle length is not actually fixed during an isometric contraction. A. V. Hill (Hill, 1938) and Katz (Katz, 1939) were among the first to suggest that sarcomere length was not constant in an isometric contraction. To overcome this confounding factor Andrew Huxley developed a servo-motor system that was capable of keeping the sarcomere length constant (Gordon, et al., 1966).

Figure 17.2 depicts the changes in muscle length that occur during an isometric contraction at the onset of muscular activation. The center sarcomere which is directly underneath the endplate is the first sarcomere to shorten (Figure 17.2 A shows the center sarcomere relaxed, and Figure 17.2 B shows the center sarcomere contracted), When the center sarcomere shortens it pulls on the sarcomeres to the right and left causing them to lengthen. This lengthening of the neighboring sarcomeres is thought to be due to passive stretch of the titin filaments. After all of the titin filaments along the muscle fiber are tight, the muscle fiber will then pull on the tendon to take up compliance in the tendon.

## SARCOMERE FORCE – LENGTH RELATIONSHIP

In the early 1960's Andrew Huxley and colleagues (Gordon, et al., 1966; Gordon, et al., 1966) and Paul Edman (Edman, 1966) used the laser diffraction technique shown above, in Figure 17.1, to determine the force-length relationship of a sarcomere. Both research groups used an isolated muscle fiber from a frog semitendinosus muscle. A typical force-length curve for a sarcomere is shown in Figure 17.3. The relationship between force and sarcomere length is proportional to the number of crossbridges that are generating tension and the amount of overlap of actin – myosin filaments. The highest force generated occurs at sarcomere lengths of 2.0 – 2.2 µm. This length is referred as optimal length as it is the length where the sarcomere can generate the greatest force. At optimal length the highest number of crossbridges can be formed so the sarcomere generates the greatest force. When the sarcomere is shortened to

lengths less than 2.0 µm, the actin – myosin filaments overlap or interfere with each other. This overlap of actin – myosin filaments causes the crossbridges to interfere with each other and as a result the force generated declines sharply. When the sarcomere is stretched to lengths greater than 2.2 µm, the motors in the center of the sarcomere cannot reach an actin binding site to form a crossbridge. As a result the force generated declines progressively with an increase in length above 2.2 µm.

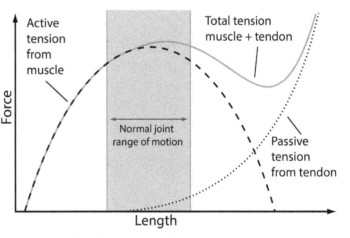

Figure 17.4 Muscle force–length and normal joint range of motion.

## FORCE – LENGTH RELATIONSHIP IN NORMAL JOINT RANGE OF MOTION

The force – length curve shown in the graph above (Figure 17.3), can be obtained from an isolated muscle fiber or whole muscle. For isolated muscle tissue it is possible to lengthen or shorten a muscle to greater or lesser muscle length than occurs in normal joint range of motion. Figure 17.4 shows the force-length curve for an isolated muscle and the shaded region depicts the limited range of motion for an intact muscle. The skeletal system and ligaments restrict motion, as a result, during normal joint range of motion the muscle is limited to a much smaller range of motion than that of an isolated muscle. The dashed line depicts the active force, which is generated by actin-myosin crossbridges. The dotted line shows passive tension from the pull of the tendon and cytoskeletal structures (titin, costamere, and desmin). The solid gray line represents the total tension or force in the muscle, it includes both the force generated by the crossbridges (dashed line) and the elastic force caused by pulling the tendon and cytoskeletal springs. The actual shape of the curve depends upon the muscle architecture.

## MUSCLE FORCE – VELOCITY RELATIONSHIP

The muscle force velocity relationship was first described in 1938 by A. V. Hill (Hill, 1938) and a year later by Katz (Katz, 1939) using an isolated frog sartorius muscle. Hill (Hill, 1938) developed the following rectangular hyperbolic equation to describe the muscle force – velocity relationship.

$$(P + a)V = b(Po - P)$$

where P is the force generated at a given velocity V, Po is maximum isometric force, and a and b are constants that are experimental derived. The rectangular hyperbola equation above has asymptotes at $P = -a$ and $V = -b$. The values for the constants a and b are typically 0.25. The constant b is in units of velocity (cm/s, mm/s) and the constant a is in units of force (N). Under zero load the maximum velocity of shortening (Vo) is defined as follows:

$$Vo = \frac{bPo}{a}$$

where Vo is the maximum velocity of shortening under zero load, Po is maximum isometric force, and a and b are constants that are experimental derived.

The force-velocity curve is shown in Figure 17.5. Inspection of the curve reveals that eccentric force > isometric force > concentric force. For concentric contractions, an increase in the velocity of shortening results in a decrease in force. In eccentric contractions an increase in the velocity of lengthening results in an increase in force. The eccentric force reaches a plateau where further increases in the velocity of stretch yield a constant eccentric force.

### Concentric Force – Velocity

The concentric force generated as a function of velocity of muscle shortening depends upon the number of myosin motors forming crossbridges and the force produced by each myosin motor. The decline in concentric force with increasing velocity of shortening is attributed to both a reduction in the myosin motor force and the number of myosin motors generating the force, (Figure 17.5). At slow velocity of shortening the myosin motor generates 6 pN of force and at the fastest velocity of shortening the force per actomyosin crossbridge decreases to 4 pN (Piazzesi, et al., 2007). In addition to this decline in motor force, increasing the velocity of shortening requires an increase in the cycling of actomyosin crossbridges, which then decreases the number of motors attached (Edman, 1988; Hill, 1938; Huxley, et al., 1971; Huxley, 2000; Piazzesi, et al., 2007). As shown in Figure 17.5, the number of motors declines from the isometric value of 80 motors at slow velocity to 20 motors at the fastest velocity of shortening (Piazzesi, et al., 2007).

**Eccentric Force – Velocity**

In an eccentric contraction increasing the velocity of muscle stretch results in an initial sharp increase in eccentric force, after which the force reaches a plateau at 1.6 – 2.0 times the maximum isometric force (Po), (Flitney & Hirst, 1978; Flitney, et al., 1978; Julian, et al., 1979; Lombardi, et al., 1990; Morgan, 1990). As show in Figure 17.5, the increase in eccentric force is attributed to an increased stiffness of the myosin motors and the cytoskeletal springs (titin, desmin, and costamere). At low velocity of stretch the motors and cytoskeletal springs increase stiffness which results in a 6 pN elastic resistive force. Increasing the velocity of stretch causes a further increase in the resistive elastic strain up to the maximum eccentric elastic strain of 14 pN (Piazzesi, et al., 2007). The myosin motors generate a maximum force of 6 pN in an eccentric contraction and the additional force is now thought to be due to non-linear stiffness of the motor and cytoskeletal structures (Edman, 2009; Nocella, et al., 2013). Colombini et al.,

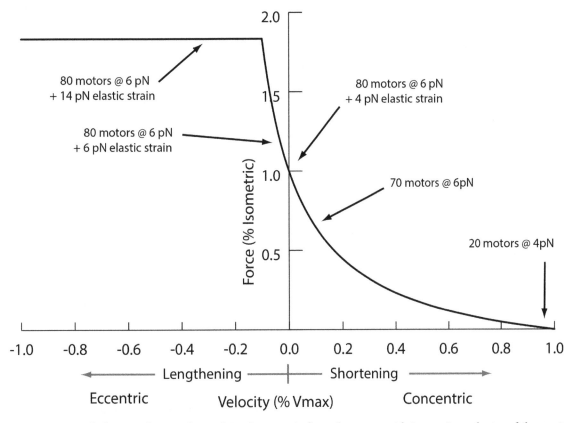

**Figure 17.5** Muscle force–velocity relationship. Concentric force decreases with increasing velocity of shortening due to a reduction in motor force and the number of attached motors. Eccentric force increases with increasing velocity of stretch due to increased elastic strain.

estimated that the force to rupture a single actomyosin crossbridge was 10.8 pN in an isometric contraction (Colombini, et al., 2007). In an eccentric contractions they found that the force required to rupture a strongly bound actomyosin crossbridge was 14 pN for slow velocity and 22 pN for fast velocity of stretch. This increased force necessary to rupture an eccentric crossbridge is thought to be due to non-linear stiffness.

Eccentric contractions are very efficient. They produce forces that are up to two times greater than isometric force at a lower metabolic cost than isometric contractions. Unlike concentric contractions where the force declines with increasing velocity, the force in eccentric contractions increases with increasing velocity. Due to non-linear stiffness of the myosin motors and cytoskeletal springs eccentric contractions produce high force at a low metabolic cost. While there may be little metabolic cost to this high resistive elastic strain, there is a mechanical cost, the mechanical strain can be destructive to the muscle tissue.

## Damage to Muscle Tissue Following Eccentric Contractions

In response to an external force that lengthens active muscle, the myosin motors and cytoskeletal springs exert a non-linear elastic force that is up to two times greater than maximum isometric force. Myosin motors and cytoskeletal springs are able to resist lengthening with forces of 11 – 22 pN (Colombini, et al., 2007) over an average distance of 11.9 nm per half sarcomere (Colombini, et al., 2009). Crossbridges are forcibly detached once this critical

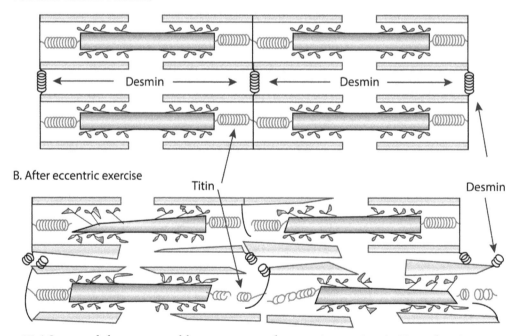

**Figure 17.6** Structural damage caused by unaccustomed eccentric exercise. A. Normal actin, myosin, titin, and desmin prior to exercise (top). B. Structural damage to actin, myosin, titin, and desmin following unaccustomed eccentric exercise (bottom).

Adapted from Allen DG. Eccentric muscle damage: mechanisms of early reduction of force. *Acta Physiologica Scandanavia*, 171: 311-319, 2001. Used with permission from John Wiley & Sons.

length of approximately 12 nm per half sarcomere is exceeded. Repeated exposure to high eccentric loading has been shown to be destructive to the muscle tissue and often leads to muscle soreness. The term DOMS (delayed onset muscle soreness) is used to describe this muscle soreness which typically develops 8 – 48 hours after unaccustomed eccentric exercise in untrained individuals, and trained individuals who do not routinely do eccentric exercise. Exercises that expose muscle tissue to higher strain forces than typically experienced may cause sufficient damage to induce DOMS.

Figure 17.6 depicts the effects of an unaccustomed bout of eccentric exercise on the muscle machinery. Following a bout of unaccustomed eccentric exercise, actin, myosin, titin, and desmin filaments show evidence of structural damage (Allen, 2001). Muscle biopsy studies have revealed that the sarcolemma is torn apart, and the elastic titin and desmin connective tissue is strained or broken (Allen, 2001; Friden, et al., 1998; Lieber, et al., 1993; Lieber, et al., 1999; Lieber, et al., 2000; Lieber, et al., 2000; Lieber, et al., 1991).

Another very unique attribute of eccentric contractions is that they primarily induce damage to fast twitch muscle fibers (Enoka, 1996; Lieber, et al., 1991). Since fast twitch fibers generate the highest force in the shortest time, they are best suited for rapidly generating the high force required to resist the eccentric loading. From a training perspective this finding has important implications for strength training and specifically power and explosive strength training. Since fast twitch fibers incur the greatest amount of tissue damage following eccentric exercise, they respond to the damage by rebuilding stronger fast twitch muscle fibers. Therefore, incorporating eccentric exercise as a normal component of muscular strength training can significantly enhance both strength and the rate of force development.

## STRETCH–SHORTEN CYCLE

A stretch-shorten cycle (SSC) is defined as an eccentric contraction followed by an immediate concentric contraction. The shortening (concentric) phase of a SSC is more powerful, does more work, than the shortening (concentric) phase of a concentric contraction (Bosco, et al., 1979; Cavagna, et al., 1974; Komi, 2000; Komi, et al., 1978; van Ingen Schenau, et al., 1997). The extra work performed during the shortening (concentric) phase of a stretch-shorten cycle is attributed to the following:

1. **Stored Elastic Energy**. During the stretch phase of the motion (eccentric), elastic structures in the muscle store elastic energy which is returned during the concentric phase of the motion. The transition between eccentric and concentric must be immediate for the elastic energy to be returned. If the transition between eccentric and concentric is not immediate the energy is dissipated within the muscle.
2. **Stretch Reflex**. The eccentric phase of the motion elicits a stretch reflex which then increases the activation of the muscle that was stretched resulting in a more forceful contraction.
3. **Preloading the Muscle**. The eccentric phase of the stretch-shorten cycle preloads the muscle. As a result, the concentric phase of the stretch-shorten cycle begins at a higher force than the concentric phase of a shorten only (concentric only) contraction.

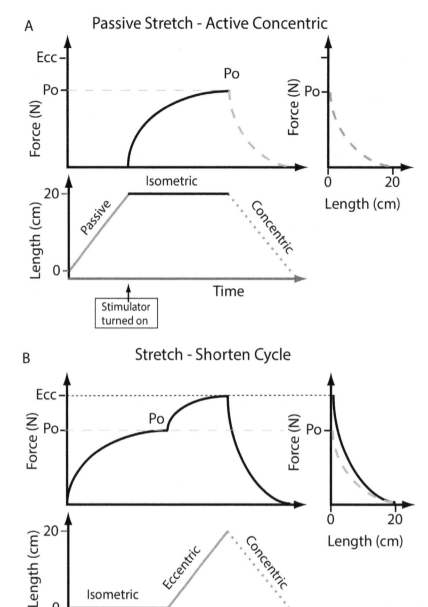

**Figure 17.7** Effects of a stretch–shorten cycle (SSC) on work done. (A) Work done in a concentric contraction. The muscle is passively pulled 20 cm, then stimulated to attain isometric max (Po), followed by a concentric contraction. The area under the force–length curve in the upper right of (A) shows the work done in the concentric phase. (B) Extra work done in a stretch-shorten cycle. The muscle is activated first, allowed to reach Po. Upon reaching Po an eccentric contraction followed by an immediate concentric contraction (SSC) is performed. The force–length curve in B illustrates the extra work done in the concentric phase of an SSC (solid line), when compared to work done in the concentric contraction (dashed line).

Adapted from Cavagna GA, Citterio G. Effect of stretching on the elastic characteristics and the contractile component of frog striated muscle. *Journal of Physiology*, 239: 1-14, 1974. Used with permission from John Wiley & Sons.

4. **Two Joint Muscles**. Two joint muscles contribute to additional work performed during a SSC. For example, during the extension phase of a vertical jump, knee and hip extension occur simultaneously. The rectus femoris, which is a two joint muscle, is shortened by the knee extension and lengthened by the hip extension. The net effect is that rectus femoris is able to generate force at a near optimal length at a lower velocity. Thus two joint muscles are able to take advantage of both muscle force – velocity and force – length relations.

## Work Done during Shortening in a Concentric Contraction

Extra work is performed during the shortening phase of the stretch-shorten cycle. To understand this effect we must begin by examining a concentric contraction to determine the work done during the shorten phase. Then we will compare this work done to the work done during the concentric phase of the stretch-shorten cycle. Figure 17.7 compares the work done in a concentric contraction to the work done in the concentric phase of a stretch-shorten cycle.

In Figure 17.7 A we will determine the work done during shortening of a concentric contraction. We will use this value to determine the extra work done in the shortening phase of a SSC. In the length – time graph at the bottom of Figure 17.7 A, the muscle is passively stretched 20 cm. Following this stretch the muscle stimulator is turned on and the muscle generates a maximum isometric contraction (Po), see the force – time graph at the top of Figure 17.7 A. As soon as the muscle reaches Po it is allow to shorten at a fixed velocity. Finally the work done is determined by computing the area underneath the force – length graph, (upper right of Figure 17.7 A).

## Work Done during Shortening in a Stretch-Shorten Cycle

The experimental conditions for Figure 17.7 B are almost identical to those in Figure 17.7 A, with one exception, before the muscle is pulled the stimulator is turned on to allow the muscle to reach maximal isometric force (Po). Thus, Figure 17.7 B shows a stretch–shorten cycle. In Figure 17.7 B the muscle is again pulled 20 cm, but since the muscle stretch begins with muscle force at Po and the muscle is active prior to the pull, we see a stair-case effect from the eccentric action. As soon as the muscle has been eccentrically stretched 20 cm it is immediately allowed to perform a concentric contraction at exactly the same velocity of shortening that was used in Figure 17.7 A. Finally, examine the force–length graph shown on the right side of Figure 17.7 B. The area underneath the solid line defines the work done during the concentric phase of the stretch–shortening in 17.7 B and the area underneath the dashed line shows the work done by the concentric only contraction from Figure 17.7 A. It is evident from the force–length curve in Figure 17.7 B that additional work was performed during the concentric phase of the stretch–shorten cycle when compared to the work done in the concentric phase of the concentric contraction. This additional work during the concentric phase of a stretch–shorten cycle is attributed to: stored elastic energy, two joint muscles, stretch reflex and preloading the muscle.

## A. Stretch - Shorten Cycle with Slow Velocity of Stretch

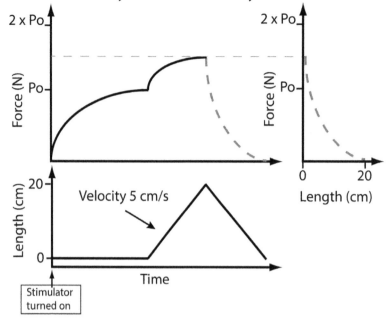

## B. Stretch - Shorten Cycle with Fast Velocity of Stretch

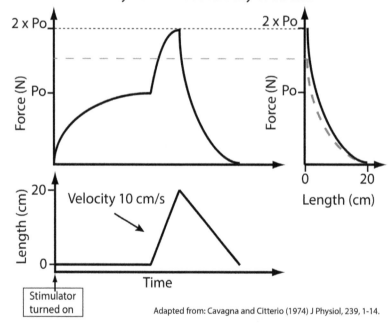

Adapted from: Cavagna and Citterio (1974) J Physiol, 239, 1-14.

**Figure 17.8** Effects of velocity of stretch on work done in a stretch–shorten cycle (SSC). (A) Slow velocity of stretch. An SSC is performed by stretching the muscle 20 cm at a slow velocity of 5 cm/s. The work done during the concentric phase is shown in the force-length graph for A, dashed line. (B) An SSC is performed by stretching the muscle 20 cm at a fast velocity of 10 cm/s. Increasing the velocity of stretch in a SSC increases the work done during the concentric phase of the SSC. The area underneath the solid line (fast velocity) is greater than the area underneath the dashed line (slow velocity), force-length graph in B.

Adapted from Cavagna GA, Citterio G. Effect of stretching on the elastic characteristics and the contractile component of frog striated muscle. *Journal of Physiology*, 239: 1-14, 1974. Used with permission from John Wiley & Sons.

## Effects of Increasing the Velocity of Stretch in a SSC

Figure 17.8 shows the effects of increasing the velocity of stretch on the work done in the concentric phase of a stretch shorten cycle. In Figure 17.8 A, the muscle is activated and allowed to reach isometric maximum force (Po). Then, in the eccentric phase, the muscle stretched a distance of 20 cm at a slower velocity (5 cm/s) followed by an immediate concentric contraction. The area underneath the force–length curve in Figure 17.8 A illustrates the work done during the concentric phase.

The experimental conditions in Figure 17.8 B are identical to those described for Figure 17.8 A, with one exception, the velocity of stretch is doubled, and thus the eccentric stretch is 20 cm at a velocity of 10 cm/s. Following the eccentric stretch an immediate concentric contraction is again performed at the same velocity of shortening as in Figure 17.8 A. The force–length curve shown on the lower right in Figure 17.8 A compares the work done during the concentric phase for the slower velocity (dashed line) to the faster velocity of stretch (solid line). Since the area underneath the solid line in the force – length curve is greater than the area underneath the dashed line, we have shown that increasing the velocity of stretch in a stretch–shorten cycle results in an increase in the work done during the concentric phase. The additional work performed following a higher velocity of stretch can be attributed to: storing additional elastic energy, increasing the pre-load on the muscle, and eliciting a greater stretch reflex.

## Stretch-Shorten Cycle in a Vertical Jump

The extra work done during the concentric phase of a stretch-shorten cycle can be demonstrated using two types of vertical jumps: squat jump (SJ), and countermovement jump (CMJ), see Figure 17.9. The squat jump represents a pure concentric contraction and the countermovement jump uses a stretch-shorten cycle. In the squat jump the subject begins by squatting down until the thighs are parallel to the ground. The subject holds this position for 2-3 seconds and then executes a maximum vertical jump by extending the hip, knee and ankle joints. This upward phase of the SJ is a concentric contraction of the hip, knee and ankle extensor muscles.

The countermovement jump is simply a normal type of vertical jump. In the CMJ, the subject rapidly lowers the center of mass by flexing the hip, knee, and ankle joints. The subject immediately reverses the downward motion, as soon as the thighs are parallel to the ground, by extending the hip, knee and ankle joints to execute a maximum vertical jump. The downward phase of the CMJ eccentrically loads the hip extensors, knee extensors and ankle plantar flexors. The upward phase of the CMJ is a concentric contraction of the hip, knee, and ankle extensor muscles. In the CMJ the hip, knee and ankle muscles all execute a stretch-shorten cycle: eccentric contraction followed by an immediate concentric contraction.

In the squat jump, shown in Figure 17.9, the subject has a vertical velocity of 1.89 m/s at takeoff and his center of mass reaches a peak of 42.46 cm above the takeoff position. In the countermovement jump the subject has a vertical velocity of 2.24 m/s at takeoff and his center of mass reaches a peak of 50.41 cm above the takeoff position. Using a stretch-shorten cycle the

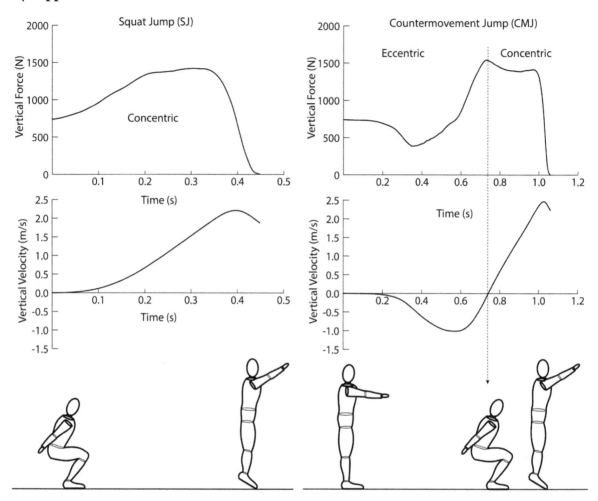

**Figure 17.9** The squat jump (SJ) begins from the squat position (shown on the left). In the SJ the hip, knee and ankle extensors undergo a concentric contraction. In the countermovement jump (CMJ) the hip, knee and ankle extensors undergo a stretch-shorten cycle, (shown on the right).

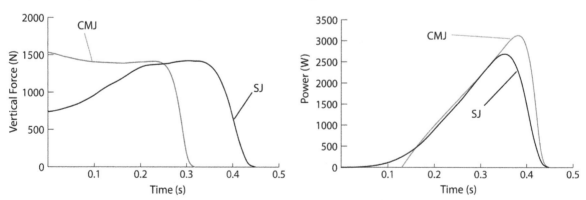

**Figure 17.10** Concentric phase of vertical force in the SJ and CMJ. In the CMJ the concentric phase begins at a higher force.

**Figure 17.11** Power – time curves for SJ and CMJ. The work done is equal to the area underneath the power-time curve. Additional work is done during the concentric phase of the CMJ when compared to the SJ.

subject was able to jump 7.95 cm higher than in the squat jump. Figure 17.10 shows the force – time curves for the concentric phases of the squat jump and the countermovement jump. At the start of the concentric phase in the squat jump the vertical force is equal to the subjects' body weight 740 N. In the countermovement jump the concentric phase begins with a vertical force of 1532 N (2 times body weight). This is an example of pre-loading the muscles. The stretch-shorten cycle pre-loads the muscles in the countermovement jump to a force of two times body weight.

The power-time curves for the concentric phase of each jump are presented in Figure 17.11. The work done in the concentric phase of each jump is equal to the area underneath the power-time curve in each jump. In the squat jump the work done is 460.58 J and in the countermovement jump the work done is 526.95 J. The extra work done in the concentric phase of the stretch-shorten cycle (CMJ) was 66.38 J. The additional 66 J of work performed during the CMJ enabled the subject to jump 7.95 cm higher when compared to the SJ.

A stretch-shorten cycle causes a more powerful muscular contraction during the concentric phase of a stretch-shorten cycle, when compared to the power production in a contraction consisting of only a concentric phase. The extra work performed during the concentric phase of a stretch-shorten cycle can be explained by the combined effects of: stored elastic energy, increased muscle activation from a stretch reflex, starting the concentric phase at a higher force, and the unique properties of two joint muscles.

## THREE COMPONENT MUSCLE MODEL

Muscles are viscoelastic tissues. As a result, muscles exhibit creep, force relaxation and they show a hysteresis. The force generated in actomyosin crossbridges elongate elastic structures (titin, desmin, costamere, and tendon) and this cytoskeletal structure must be moved through a fluid environment. Similar to all viscoelastic tissues it will be stiffer when loaded fast than when loaded slowly. The concept of muscle viscosity was first proposed by A. V. Hill (Hill, 1922) to explain the decline in force with increasing velocity of shortening. Over a period of several years Hill further investigated the viscoelastic response of muscle tissue (Hill, 1926; Hill, 1938; Hill, 1949; Hill, 1950; Hill, 1953). Hill developed a three-component muscle model to describe the mechanical response to muscle and explain the viscoelastic effect of muscle. One of the earliest published pictorial versions of the three component model by Jewell and Wilkie includes both the SEC, CC and PEC, but does not include the dashpot for viscous damping (Jewell, et al., 1958).

Figure 17.14 shows an example of a three component muscle model. The three component muscle model describes the mechanical response of muscle. The mechanical model includes contractile, series elastic and parallel elastic components. The contractile component (CC) models the behavior of the active crossbridges generating tension. The series elastic component (SEC) models the viscoelastic response of muscle, it includes a dashpot to model the viscous damping in series with a spring to model elastic structures of muscle. The anatomical structures included in the series elastic component are the tendon, titin, desmin, costamere, and the bending or stretching of the myosin arms. The tendon is the primary elastic structure modeled in the SEC. The final component is the parallel elastic component (PEC) includes passive elastic structures: tendon, passive crossbridges, titin and desmin. The parallel elastic component primarily detracts or takes away from the force generated by active crossbridges. Examples of the PEC include the dragging of passive muscle fibers during a contraction and the resistance of passive antagonists during the shortening of a muscle.

The series elastic component has been studied extensively, it explains the stored elastic energy in a stretch-shorten cycle. During the eccentric phase of a SSC energy is stored in the tendon and other elastic structures which is then returned during the shortening (concentric) phase of the SSC. In addition, the SEC explains the delay between the electrical activation of the muscle and the onset of tension in the muscle.

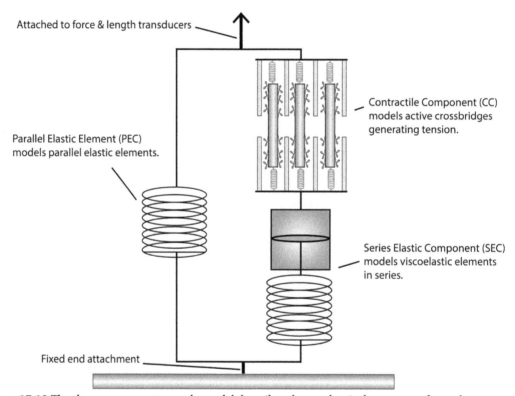

**Figure 17.12** The three component muscle model describes the mechanical response of muscle.

## REVIEW QUESTIONS

1. Draw and explain the force-length curve for a sarcomere.
2. Draw and explain the force-velocity curve.
3. Define a stretch-shorten cycle and give an example of a movement that utilizes a stretch-shorten cycle.
4. List four explanations for the additional work that is done during the concentric phase of a stretch-shorten cycle.

5. List the anatomical structures that are associated with each of the following components of the three component muscle model: parallel elastic component, contractile component, and series elastic component.

## References

Allen DG. Eccentric muscle damage: Mechanisms of early reduction of force. Acta Physiologica Scandanavia. 171: 311-319, 2001.

Bosco C, Komi PV. Potentiation of the mechanical behavior of the human skeletal muscle through prestretching. Acta Physiologica Scandanavia. 106: 467-472, 1979.

Cavagna GA, Citterio G. Effect of stretching on the elastic characteristics and the contractile component of frog striated muscle. Journal of Physiology. 239: 1-14, 1974.

Colombini B, Bagni MA, Romano G, Cecchi G. Characterization of actomyosin bond properties in intact skeletal muscle by force spectroscopy. Proc Natl Acad Sci U S A. 104: 9284-9289, 2007.

Colombini B, Nocella M, Benelli G, Cecchi G, Griffiths PJ, Bagni MA. Reversal of the myosin power stroke induced by fast stretching of intact skeletal muscle fibers. Biophysical Journal. 97: 2922-2929, 2009.

Edman KA. Double-hyperbolic force-velocity relation in frog muscle fibres. J Physiol. 404: 301-321, 1988.

Edman KA. The relation between sarcomere length and active tension in isolated semitendinosus fibres of the frog. Journal of Physiology. 183: 407-417, 1966.

Edman KAP. Non-linear myofilament elasticity in frog intact muscle fibres. J Exp Biol. 212: 1115-1119, 2009.

Enoka RM. Eccentric contractions require unique activation strategies by the nervous system. J Appl Physiol. 81: 2339-2346., 1996.

Flitney FW, Hirst DG. Cross-bridge detachment and sarcomere 'give' during stretch of active frog's muscle. Journal of Physiology. 276: 449-465, 1978.

Flitney FW, Hirst DG. Filament sliding and energy absorbed by the cross-bridge in active muscle subjected to cyical length changes. Journal of Physiology. 276: 467-479, 1978.

Friden J, Lieber RL. Segmental muscle fiber lesions after repetitive eccentric contractions. Cell Tissue Res. 293: 165-171, 1998.

Gordon AM, Huxley AF, Julian FJ. Tension development in highly stretched vertebrate muscle fibres. The Journal of Physiology. 184: 143-169, 1966.

Gordon AM, Huxley AF, Julian FJ. The variation in isometric tension with sarcomere length in vertebrate muscle fibres. The Journal of Physiology. 184: 170-192, 1966.

Hill AV. The maximum work and mechanical efficiency of human muscles, and their most economical speed. Journal of Physiology. 56: 19-41, 1922.

Hill AV. The viscous elastic properties of smooth muscle. Proceedings of the Royal Society of London Series B, Containing Papers of a Biological Character. 100: 108-115, 1926.

Hill AV. The heat of shortening and the dynamic constants of muscle. Proceedings of the Royal Society of London Series B, Biological Sciences. 117: 136-195, 1938.

Hill AV. The abrupt transition from rest to activity in muscle. Proceedings of the Royal Society of London Series B, Biological Sciences. 136: 399-420, 1949.

Hill AV. The series elastic component of muscle. Proceedings of the Royal Society of London Series B, Biological Sciences. 137: 273-280, 1950.

Hill AV. The mechanics of active muscle. Proceedings of the Royal Society of London Series B, Biological Sciences. 141: 104-117, 1953.

Hill DK. The optical properties of resting striated muscle; the effect of rapid stretch on the scattering and diffraction of light. J Physiol. 119: 489-500, 1953.

Huxley AF. Muscle structure and theories of contraction. Progress in Biophysics and Biophysical Chemistry. 7: 255-318, 1957.

Huxley AF, Simmons RM. Proposed mechanism of force generation in striated muscle. Nature. 233: 533-538, 1971.

Huxley HE. Past, present and future experiments on muscle. Philos Trans R Soc Lond B Biol Sci. 355: 539-543, 2000.

Jewell BR, Wilkie DR. An analysis of the mechanical components in frog's striated muscle. The Journal of Physiology. 143: 515-540, 1958.

Julian FJ, Morgan DL. The effect on tension of non-uniform distribution of length changes applied to frog muscle fibres. Journal of Physiology. 293: 379-392, 1979.

Katz B. The relation between force and speed in muscular contraction J Physiol. 96: 45-64, 1939.

Komi PV. Stretch-shortening cycle: A powerful model to study normal and fatigued muscle. Journal of Biomechanics. 33: 1197-1206, 2000.

Komi PV, Bosco B. Utilization of stored elastic energy in leg extensor muscles by men and women. Medicine and Science in Sports. 10: 261-265, 1978.

Lieber RL, Friden J. Muscle damage is not a function of muscle force but active muscle strain. Journal of Applied Physiology. 74: 520-526, 1993.

Lieber RL, Friden J. Mechanisms of muscle injury after eccentric contraction. J Sci Med Sport. 2: 253-265, 1999.

Lieber RL, Friden J. Functional and clinical significance of skeletal muscle architecture. Muscle Nerve. 23: 1647-1666, 2000.

Lieber RL, Leonard ME, Brown-Maupin CG. Effects of muscle contraction on the load-strain properties of frog aponeurosis and tendon. Cells Tissues Organs. 166: 48-54, 2000.

Lieber RL, Woodburn TM, Friden J. Muscle damage induced by eccentric contractions of 25% strain. J Appl Physiol. 70: 2498-2507, 1991.

Lombardi V, Piazzesi G. The contractile response during steady lengthening of stimulated frog muscle fibres. J Physiol. 431: 141-171, 1990.

Morgan DL. New insights into the behavior of muscle during active lengthening. Biophysical Journal. 57: 209-221, 1990.

Nocella M, Bagni M, Cecchi G, Colombini B. Mechanism of force enhancement during stretching of skeletal muscle fibres investigated by high time-resolved stiffness measurements. Journal of Muscle Research and Cell Motility. 34: 71-81, 2013.

Piazzesi G, Reconditi M, Linari M, Lucii L, Bianco P, Brunello E, et al. Skeletal muscle performance determined by modulation of number of myosin motors rather than motor force or stroke size. Cell. 131: 784-795, 2007.

van Ingen Schenau GJ, Bobbert MF, de Haan A. Does elastic energy enhance work and efficiency in the stretch-shortening cycle? Journal of Applied Biomechanics. 13: 389-415, 1997.

## Suggested Supplemental Resources

Enoka, R. M. (2008). Neuromechanics of Human Movement. Champaign, Human Kinetics.

Hall, S. J. (2007). Basic Biomechanics. Boston, McGraw Hill.

Hamill, J. and K. M. Knutzen (2009). Biomechanical Basis of Human Movement. Philadelphia, Lippincott Williams & Wilkins.

Kandel, E. R., J. H. Schwartz, et al. (2013). Principles of Neural Science. New York, McGraw-Hill.

Lieber, R. L. (1992). Skeletal Muscle Structure and Function. Implications for Rehabilitation and Sports Medicine. Baltimore, Williams & Wilkins.

MacIntosh, B. R., P. F. Gardiner, A. J. McComas. (2006). Skeletal Muscle Form and Function, Champaign, Human Kinetics.

McGinnis, P. M. (2013). Biomechanics of Sport and Exercise. Champaign, Human Kinetics.

Pollack G. H. (1990). Muscles & Molecules. Uncovering the Principles of Biological Motion. Seattle, Ebner & Sons Publishers.

Winter, D. A. (2009). Biomechanics and Motor Control of Human Movement. Hoboken, John Wiley & Sons, Inc.

# Chapter 18

## EMG AND MOTOR NEURONS

### STUDENT LEARNING OUTCOMES

**After reading Chapter 18, you should be able to:**

- Define a motor unit.
- Explain the difference between a muscle fiber action potential and a motor unit action potential.
- Explain the relationship between muscle fiber action potentials and the EMG signal.
- Describe a motor unit twitch response.
- Explain the effects of stimulation/firing rate on force production.
- Describe fiber type and firing rate-force properties for motor units.
- Explain how synchronization of motor units causes amplification and cancellation.
- Describe motor unit recruitment and firing rate for a ramp and ballistic contraction.
- Explain how rate of force development alters the size principle.
- Define and explain electromechanical delay.
- Describe the concentric and eccentric EMG-force relationship.

### MOTOR UNIT

A motor unit is one alpha motor neuron and all of the muscle fibers that it innervates. As shown in Figure 18.1, the cell body for the motor neuron is located in the spinal cord and the alpha nerve fiber extends from the spinal cord to the innervated muscle tissue. The nerve fiber terminates at the endplate, which is located adjacent to the muscle fibers. The number

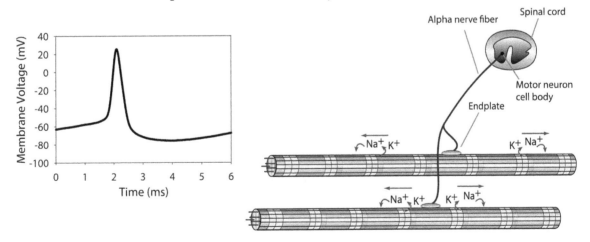

**Figure 18.1** A motor unit is one alpha motor neuron and all of the fibers it innervates. Stimulation of the nerve leads to the generation of a muscle fiber action potential (left). The muscle fiber action potential is propagated in each direction along the muscle fiber. The EMG signal arises from the summation of all the muscle fiber action potentials.

of muscle fibers controlled by a single motor neuron can vary from 10–800. The activation of muscle fibers belonging to an individual motor unit is guided by the "all or none principle". Thus, when a single stimulus is propagated down the alpha motor neuron, it will cause all of the muscle fibers belonging to that motor neuron to generate tension. As a result, a motor unit that controls a smaller number of muscle fibers will produce a smaller twitch force than a motor unit controlling a larger number of fibers. In muscles where fine–detailed movements and force gradations are required, a motor unit controls a small number of muscle fibers. For example, in finger muscles such as the first dorsal interosseus (FDI) and the ocular muscles that move the eye, a single motor unit will control as few as 10–20 muscle fibers, which then enables the fine–detailed control of motion and force in these muscles. At the other end of the continuum, in large muscles such as the vastus lateralis or latissimus dorsi, an individual motor unit generally controls 500–800 muscle fibers. Therefore, when a motor unit in the vastus lateralis is stimulated, it will generate a relatively large increment of muscular tension. The muscle fibers for a given motor unit are generally confined to a localized region of the muscle, but in that region there are fibers from several other motor units. The number of motor units in a muscle also is variable, from as few as 12 to as many as 4000.

Stimulation of the alpha motor neuron either at the spinal cord level or somewhere along the nerve causes a stimulus to be propagated down the nerve. When the stimulus arrives at the endplate, the neurotransmitter acetylcholine (ACh) is released from the endplate. Acetylcholine diffuses across the synapse which then causes $Na^+$ and $K^+$ channels in the muscle membrane to open up. In response to the opening of the membrane ion channels, $Na^+$ flows into the cell and $K^+$ flows out of the cell. This exchange of $Na^+$ and $K^+$ ions across the muscle fiber membrane results in the generation of a muscle fiber action potential (shown in the upper left of Figure 18.1). The flow of $Na^+$ into the cell causes the membrane potential be positive. The sodium channels close and the pumping of K+ out of the muscle cell returns the membrane back to its negative resting potential. The EMG signal is produced by the summation of muscle fiber action potentials which are propagated thru muscle tissue.

The muscle fiber action potential is propagated in each direction along the muscle fiber. When an individual sarcomere receives the muscle fiber action potential, it then undergoes the steps to generate tension in the fiber by forming crossbridges between actin and myosin. The initial tension generated in the muscle fiber causes the affected sarcomere to shorten and the adjacent sarcomeres are elongated (stretch of titin). Once the entire fiber length has been tightened, it will then pull on the tendon, and consequently, force can be measured at the tendon. Each stimulus that is propagated down the motor neuron will generate a muscle fiber action potential and contribute to the force generated by the muscle.

## DEFINITION OF THE ELECTROMYOGRAPHY SIGNAL

The **electromyogram (EMG)** is defined as the sum of all of the muscle fiber action potentials from all of the active motor units that pass thru the recording zone of the electrodes. The relationship between the muscle fiber action potentials from a single motor unit and the EMG signal recorded by an EMG amplifier is shown in Figure 18.2. The EMG signal can be recorded by placing electrodes on the skin surface, inserting a needle electrode into the muscle, or inserting a fine–wire electrodes into the muscle. The image on lower left hand side of Figure 18.2 shows surface electrodes which have been placed on the skin surface over the bicep muscle. The electrodes are then connected to an EMG amplifier. The illustration in the upper left of Figure 18.2 depicts an enlargement showing the skin surface, the electrodes, and a single motor unit along with three of the muscle fibers that it innervates. The shaded region underneath the electrodes depicts the recording zone of the electrodes. EMG electrodes can detect muscle fiber action potentials from some volume of muscle tissue. The size of the

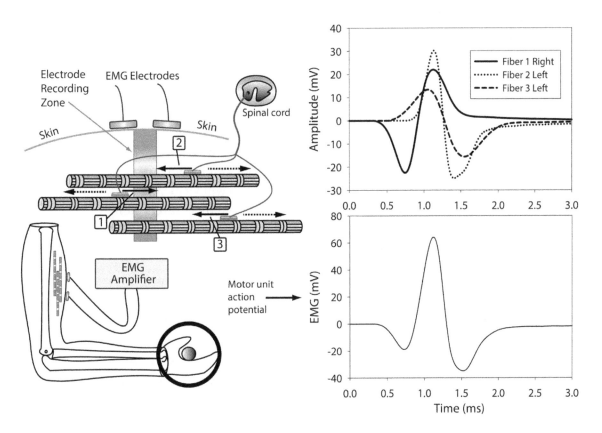

**Figure 18.2** Relationship between muscle fiber action potentials (upper right) and motor unit action potential (lower right). The electrodes (upper left) record all muscle fiber action potentials that pass thru the recording zone. Fiber potentials with dashed arrows do not pass thru the recording zone and are not recorded. The EMG amplifier (lower left) takes the difference between the fiber potentials recorded from each electrode. The muscle fiber potential from fiber 1 propagated to the right is the first to reach the recording zone. After taking the difference between the two electrodes, the resulting signal has an initial negative value (Fiber 1 Right). The muscle fiber potential traveling to the left from fiber 2 is the next to reach the recording zone. After taking the difference between each electrode signal, the muscle fiber potential for fiber 2 has an initial positive value (Fiber 2 Left). Finally, the muscle fiber potential traveling to the left from fiber 3 reaches the recording zone. After taking the difference between each electrode signal, the muscle fiber potential for fiber 3 has an initial positive value (Fiber 3 Left). The EMG signal is composed of the sum of all the muscle fiber action potentials that that pass thru the recording zone of the electrodes. The summation of muscle fiber action potentials 1-3 results in the motor unit action potential shown in the bottom right. This motor unit action potential is the signal that the EMG amplifier would record after the motor unit fires once in response to a stimulus.

recording volume of the electrodes depends upon the muscle fiber characteristics, electrode: type, spacing, and alignment relative to the muscle fibers. Once a stimulus is propagated down the motor neuron, it arrives at the terminal branches of each muscle fiber innervated by the motor unit. Muscle fiber action potentials are subsequently propagated in each direction along the muscle fibers as described earlier in Figure 18.1. The EMG electrodes will record all muscle fiber action potentials that pass thru the recording zone of the electrodes. The muscle fiber action potentials shown in the upper right hand side of Figure 18.2 represent the fiber potentials recorded by the EMG electrodes from muscle fibers 1, 2, and, 3, respectively. Finally, the motor unit action potential shown in the lower right hand corner of Figure 18.2 is the sum of the individual muscle fiber action potentials recorded from fibers 1, 2, and 3. The relationship between muscle fiber action potentials and the recorded EMG signal is affected by the following:

1. The direction the muscle fiber action potential travels thru the recording zone determines its polarity (whether it is positive first or negative first).
2. The distance from the recording electrodes to the active fibers affects the amplitude of the recorded signal. Fibers that are closest to the electrodes will yield large amplitude signals, whereas fibers that are further away from the electrodes will yield small amplitude signals.
3. The horizontal distance that the muscle fiber action potential travels along in the muscle fiber affects the time of recording the signal.

**Direction Determines Polarity**

Three muscle fibers from a single motor unit are shown in Figure 18.2. For each fiber, the dashed line arrows depict muscle fiber action potentials that are propagated away from the recording zone and therefore will not be recorded by the electrodes and EMG amplifier. The solid arrows on each muscle fiber show fiber action potentials that pass thru the recording zone, which will then contribute to the resulting motor unit action potential.

The endplate for muscle fiber 1 in the Figure 18.2 is closest to the recording zone and as a result, it will be the first fiber potential recorded by the electrodes. For fiber 1, the muscle fiber action potential propagated to the left (dashed line arrow) does not pass thru the recording zone, so it will not be recorded. The signal that is propagated to the right (solid arrow) does pass thru the recording zone, so it will be recorded. The EMG amplifier takes the difference between the muscle fiber action potential recorded by each electrode. The resulting signal is shown to right of the muscle fiber (muscle fiber action potential 1). For fiber 1, the muscle fiber action potential travels to the right as it passes thru the recording zone. Once the amplifier takes the difference between the signal captured by each electrode the resulting signal will be negative first then positive.

The endplate for fiber 2 is closer to the recording zone than the endplate for fiber 3. As a result, fiber 2 will be the next muscle fiber action potential to be recorded by the electrodes. For fiber 2, the signal that is propagated to the right does not pass thru the recording zone (dashed line arrow) and the signal that is propagated to the left does pass thru the recording

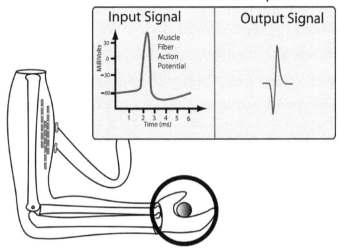

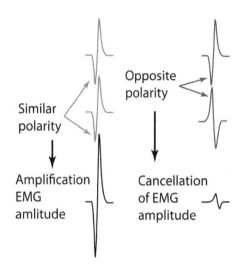

**Figure 18.3** The EMG signal is recorded using a differential amplifier. The amplifier takes the difference between the muscle fiber action potential recorded from each electrode as input and it outputs the EMG signal shown in the right.

**Figure 18.4** Synchronization of motor unit firing causes amplification and/or cancellation of EMG amplitude.

zone. The resulting muscle fiber action potential recorded from fiber 2 will be positive first, then negative (EMG signal shown to the right, number 2).

Finally, for fiber 3, which is the furthest away from the recording zone, the signal that is propagated to the left (solid arrow) passes thru zone and just as in signal 2, the muscle fiber action potential from fiber 3 has a polarity that is initially positive and then negative (muscle fiber action potential number 3).

### Distance Affects Amplitude

An EMG amplifier takes the difference between a muscle fiber action potential recorded by each electrode. Figure 18.3 illustrates the input signal from each electrode and the resulting output signal from the EMG amplifier. At the muscle fiber level, the muscle fiber action potentials arising from a single motor unit are very similar to each other and subsequently, they look like the signal in Figure 18.1. The amplitude of an extracellular potential is attenuated as it passes thru tissue to reach the recording electrodes. In Figure 18.2, fiber 2 has the smallest vertical distance from the fiber to the skin surface where the recording electrodes are located whereas fiber 3 has the greatest vertical distance. Since the distance attenuates the signal amplitude, fiber 2 has the highest peak to peak amplitude and fiber 3 has the smallest peak to peak amplitude.

### Horizontal Distance along the Muscle Fiber Determines the Time of Recording

The muscle fiber action potential is propagated longitudinally along the muscle fiber at 3–5 m/s. Since the endplate for fiber 1 is closest to the recording zone, it is recorded in time first, followed by fiber 2, and then fiber 3. The summation of the three fiber potentials shown in the upper right of Figure 18.2 produces the motor unit action potential shown in the lower right of Figure 18.2. This motor unit action potential is the signal that would be recorded after delivering a single stimulus to the nerve with a muscle stimulator.

### Synchronization Causes Amplification and Cancellation of Fiber Potentials

Synchronization occurs when two or more motor units (MU) fire or are stimulated at the same time. Figure 18.4 illustrates how synchronization affects the amplitude of the recorded EMG signal. The left side of the Figure 18.4 shows the resulting amplification of EMG amplitude when two synchronized motor units with similar polarity are summed. Since the two motor units shown on the left have similar polarity (both are negative first, then positive), the addition of the two signals results in a greatly amplified EMG signal. On the right side of Figure 18.4, the synchronized motor units have opposite polarity and they effectively cancel each other out, resulting in a very small amplitude EMG signal. Thus, synchronization can both increase and/or decrease the amplitude of the EMG signal depending upon the time of recording and polarity of the underlying muscle fiber action potentials.

## MUSCLE CONTRACTILE RESPONSE

### Twitch Response

Figure 18.5 shows a muscle twitch response. A muscle twitch response is the force–time curve that is generated when a single stimulus causes a motor unit to produce muscle tension. The magnitude and shape of a twitch response is characterized by: the peak force, time to peak force, and the time for the force to decline ½ of the peak value also referred to as one-half relaxation time. The magnitude and time to peak force are used to delineate fiber types from a twitch response. Slow twitch fibers have the longest time to peak force and lowest magnitude of peak force. Intermediate muscle fibers have a shorter time to peak and a higher peak force than slow twitch fibers. Fast twitch muscle fibers generate highest magnitude of twitch force in the shortest time, when compared to slow or intermediate muscle fibers.

### Stimulation Rate and Tetanic Contractions

Figure 18.6 shows the "stair-case" effect of stimulation rate on the muscle force–time relationship. The muscle force–time curve shown in Figure 18.6-A is a twitch contraction produced by a single stimulus where the muscle is allowed to return to full relaxation before additional stimuli are delivered. The force–time curve shown in Figure 18.6-B illustrates the effect of increasing the rate of stimulation on the resulting muscle force output. In Figure 18.6-B, three stimuli are delivered at a rate approximately equal to the time to maximum force. Each subsequent stimulus causes an increase in the force response when compared to the previous stimulus. Finally, in Figure 18.6-C, the force–time curve represents a tetanic contraction. A tetanic contraction results by increasing the rate of stimulation until no further increase in the force response occurs. Typically, a slow twitch motor unit will reach tetanic force at stimulation rates of 60–80 Hz (times per second), an intermediate unit will reach tetanic force at 60–90 Hz and a fast twitch motor unit reaches tetanic force at 80–120 Hz. Increasing the rate

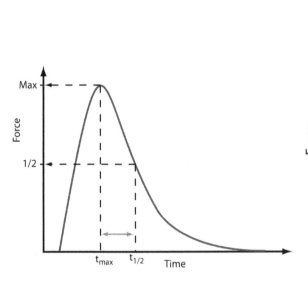

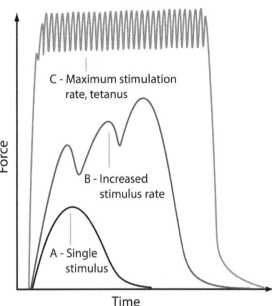

**Figure 18.5** A muscle twitch response is the force–time curve generated when a stimulation signal causes a motor unit to fire once.

**Figure 18.6** Increasing the stimulation rate of a muscle produces a "stair-case" effect of muscle force. **A** is the force-time curve from a single stimulus. **B** shows the effect of increasing the stimulation rate on the force output. In **C,** the maximal stimulation rate yields a tetanic contraction.

of stimulation causes an increase in the force generated by a muscle and the rate of force development. Comparison of the time to maximum force for the three force – time curves in Figure 18.6 shows that the tetanic stimulation results in a much higher rate of force development.

The central nervous system (CNS) can change the level of muscle force by increasing the stimulation rate and/or recruiting additional motor units. The stimulation rate is also referred to as the firing rate of a motor unit. Any combination of recruitment of additional motor units and or increasing the firing rate of currently active motor units will result in an increase in muscle force output.

## THE SIZE PRINCIPLE AND MOTOR UNIT PROPERTIES

In 1957, Elwood Henneman (Henneman, 1957) observed that motor units seemed to be recruited according to the size of the motor neuron. Over the course of several additional papers, Henneman (Henneman, 1985; Henneman, et al., 1974; Henneman, et al., 1965; Henneman, et al., 1965; Henneman, et al., 1965) formally stated what is commonly referred to as the size principle of motor unit recruitment. According to the size principle, in a contraction that progresses from low force to maximum force, small, slow twitch units (ST) are recruited first, followed by intermediate motor units (INT), and finally large, fast twitch (FT) units are the last to be recruited. The motor units are turned off (derecruited) in reverse order when the muscle force progresses from maximum force to relaxation, meaning fast twitch are turned off first, followed by intermediate, and finally slow twitch motor units.

Figure 18.7 shows the relationship between the EMG signal, fiber type, firing rate (stimulation rate), nerve diameter, nerve conduction velocity, twitch response, and fatigability. The size principle can be generalized to include the diameter of the nerve, size of the EMG amplitude, and the size of the twitch force response from the motor unit. In other words, in a muscle contraction where the force slowly increases, small diameter nerves that generate small amplitude EMG and force are the first to be recruited and large diameter nerves that generate high amplitude EMG and force are the last to be recruited.

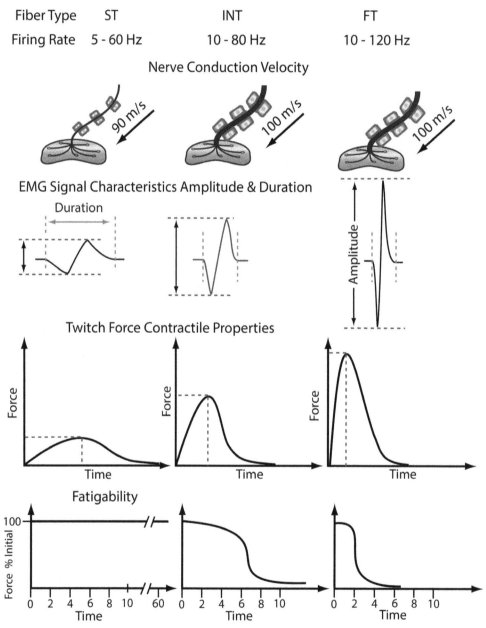

**Figure 18.7** Relationship between EMG signal properties and muscle fiber characteristics, twitch force and fatigability.

## Nerve Conduction Velocity

The nerve for a slow twitch motor unit has less myelin than nerves belonging to intermediate and fast twitch motor units. Since myelin is an insulator, the slow twitch neuron has a slower nerve conduction velocity than the intermediate and fast twitch fibers. As shown in Figure 18.7, nerve conduction velocity for slow twitch motor units is 90 m/s. In comparison, nerve conduction velocity is 100 m/s for intermediate and fast twitch motor units.

## Motor Unit Firing Rate

The stimulation rate for a motor unit is also referred to as its firing rate or discharge rate. All three of these terms are usually given in units of Hz (times per second). The firing rate range for slow twitch motor units is 5–60 Hz. Intermediate motor units have firing or discharge rates of 10–80 Hz. Finally. fast twitch motor units have firing rates of 10–120 Hz (see Figure 18.7).

## EMG Signal Characteristics

The EMG signal for a single stimulus that elicits a twitch response is characterized by the amplitude and duration of the motor unit's action potential. As shown in Figure 18.7, a slow twitch unit has relatively small amplitude and longer duration. The motor unit action potential from an intermediate motor unit has higher amplitude and shorter duration than a slow twitch motor unit. Finally the fast twitch motor unit action potential has the highest amplitude and the shortest duration of the three fiber types.

## Twitch Force Characteristics

The twitch force from a slow twitch motor unit has the lowest peak force and the longest time to peak tension. The twitch force from an intermediate motor unit gives higher peak force with a shorter time to peak force when compared to a slow twitch motor unit. Finally, the fast twitch motor unit generates the highest magnitude of peak force and has the smallest time to peak of the three fiber types.

## Fatigability of Motor Unit Types

Slow twitch motor units are very resistant to fatigue; they can be driven (stimulated) continuously for 60 minutes or longer. Depending upon the level of force, intermediate motor units can be driven continuously for 4–6 minutes and they can fire sporadically for 60 minutes or more. Fast twitch motor units can be driven continuously for about 2 minutes and they will fire sporadically beyond two minutes.

## SUMMARY OF MOTOR UNIT PROPERTIES

According to the size principle, motor units are recruited according to size: small, then intermediate, then fast. Motor units are derecruited in reverse order: fast twitch units are turned off first, then intermediate, and finally slow twitch units. As a result, slow twitch units will be utilized throughout the entire contraction, whereas fast twitch units are only activated at high force levels, or when a high rate of force development is required. The size principle

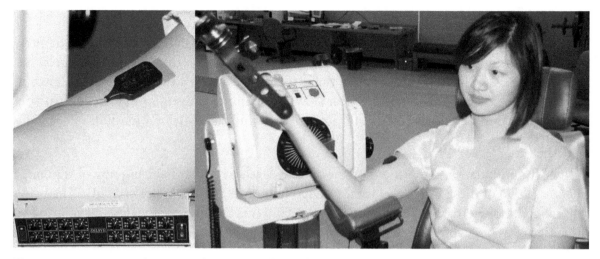

**Figure 18.8** Experimental setup used to measure biceps brachii EMG and elbow torque for an isometric ramp and ballistic contractions. An EMG electrode is placed over the bicep muscle (upper left) and then attached to an EMG amplifier (lower left). The subject generates an isometric contraction and the elbow torque is measured with an isokinetic strength machine.

can be generalized to describe the diameter of the nerve, amplitude of the EMG signal, and the amplitude of the peak twitch force. Fast twitch motor units have the largest diameter nerve, highest maximal firing rate, highest amplitude EMG signal, and the highest magnitude of peak force with the shortest time to peak tension. Slow twitch motor units have the smallest diameter nerve fiber, the lowest maximal firing rate, the lowest amplitude of EMG signal, and the lowest magnitude of peak force with the longest time to peak tension. Fatigability of motor units is inversely related to their recruitment order. Slow twitch motor units are very fatigue resistant and as a result, they are the first to be recruited and the last motor units to be derecruited in a contraction. Fast twitch motor units are highly fatigable, and as a result, the central nervous system reserves recruitment of fast twitch units until high force and/or high velocity is required in a contraction.

## FORCE MODULATION WITH MOTOR UNIT RECRUITMENT AND FIRING RATE

Muscles have two mechanisms available to alter or modulate the level of force that is generated: recruitment and firing rate. The force generated by a muscle can be increased by increasing the firing rate of currently active motor units, recruiting additional motor units, or a combination of both increasing firing rate and recruitment of new motor units. An increase in the discharge rate of a motor unit will cause a "stair-step" effect of increasing force. Recruiting additional motor units will also increase the force generated. Figure 18.8 illustrates the typical experimental setup to measure EMG during an isometric contraction. The torque generated is measured using an isokinetic device or a load cell. The EMG signal is measured using either surface or needle electrodes. Using surface electrodes, as shown in Figure 18.8, will not allow for the tracking of individual motor units to determine both recruitment and firing rate as a function of torque. Needle electrodes are required to actually track individual motor units.

## Ramp Contraction

A ramp contraction is an isometric contraction where the subject slowly and smoothly increases the force in a linear fashion from 0 to 100% of a maximum voluntary contraction (MVC). The duration of a ramp contraction typically takes 2–10 seconds to go from 0 to 100% MVC. Figure 18.9 shows motor unit recruitment and firing rate changes and the resulting

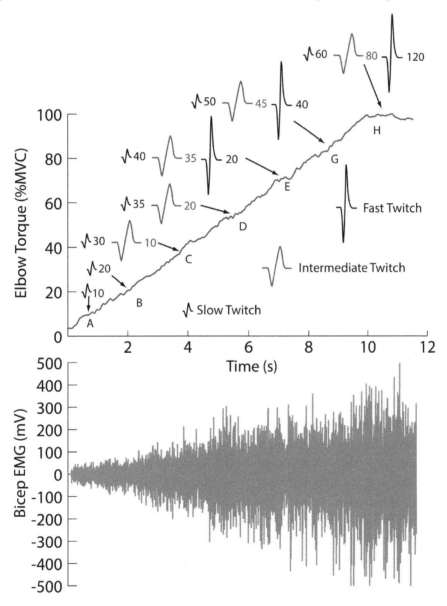

**Figure 18.9** Relationship between motor unit recruitment and firing rate changes and the resulting linear increase in force and EMG amplitude for a ramp contraction. The graph on the top shows the relationship between EMG recruitment and firing rate. Individual motor units are shown adjacent to the force–time curve, with small peak to peak amplitude for ST, intermediate amplitude for INT, and large amplitude for FT. The numbers beside the motor unit action potentials represent the firing rate for each motor unit type.

EMG signal for a ramp contraction. The force–time curve for the ramp contraction is shown in the top graph of Figure 18.9 and the bottom graph displays the simultaneous EMG–time curve for the ramp contraction. The force is presented in units of maximum voluntary contraction (MVC). A maximum voluntary contraction is the maximal force that an individual can generate. The motor unit potentials shown to the left of the force curve represent the fiber type for a given motor unit. Small peak–to–peak amplitude represents a slow twitch motor unit, intermediate amplitude represents an intermediate motor unit, and the largest amplitude represents a fast twitch motor unit. The corresponding numbers beside each motor unit signal is the motor unit's firing rate.

Figure 18.9 shows the size principle in action. At the start of the contraction, the force is initially provided by slow twitch motor units with a firing rate of 10 Hz (labeled with A on the force curve). Then at time point B, the currently active slow twitch motor units increase firing rate to 20 Hz. The increase in force from A to B is only due to changes in firing rate. At time point C, an intermediate motor unit is recruited and the slow twitch motor unit has increased its firing rate to 30 Hz. The increase in force from B to C is due to both recruitment and firing rate changes. Point C also displays a principle that occurs in a ramp contraction: **when a new motor unit is recruited, the currently active units are discharging (firing) at a higher rate than the newly recruited motor unit**. At time point D, both the slow and intermediate motor units increase their firing rates with the slow twitch units discharging are 35 Hz and the intermediate units discharging at 20 Hz. The increase in force from C to D is attributed solely to changes in the firing rate of currently active motor units. At time point E, a fast twitch motor unit is recruited and it is activated at a discharge rate of 10 Hz. Again, when a new unit is recruited, the currently active units are firing at a higher rate. For point D, the currently active units fire at 40 Hz and 35 Hz for the slow and intermediate units, respectively. At point G, all three motor units increase their firing rate. Finally, at time point H, all three motor unit types are now firing at their maximal rate: slow twitch at 60 Hz, intermediate at 80 Hz and fast twitch at 120 Hz. Point H represents is the maximal voluntary contraction force for the subject. In an MVC, all of the motor units for the muscle are recruited and they all fire (discharge) at their maximal rate. The corresponding EMG–time curve shown on the bottom of the figure shows a linear increase in the amplitude of the EMG signal during this ramp contraction. An increase of firing rate and/or a recruitment of additional motor units will cause the amplitude of the EMG signal to increase.

The derecruitment curve for this ramp contraction is analogous to playing the force–time, recruitment, and firing rate changes backwards. From H to E, the motor units would decrease their firing rates. As soon as the force goes below the force at point E, the fast twitch motor unit would shut down. From E to C, the slow twitch and intermediate units would decrease their firing rates. Once the force goes below the level at C, the intermediate unit would be derecruited (shut down). Finally from C to A, the slow twitch motor unit would decrease its firing rate in response to the reduce requirement for force.

## Force Threshold

The force at which a motor unit is recruited is called its force threshold or recruitment threshold. Motor units are recruited and derecruited at their force threshold. Point C in the ramp contraction shown in Figure 18.9 is the recruitment threshold for the intermediate unit. When the required force is at the threshold, the intermediate unit is recruited and as soon as the required force goes below the force threshold, the intermediate unit is derecruited.

## Ballistic Contraction

In an isometric ballistic contraction, the subject is instructed to generate maximum voluntary force (MVC) as rapidly as possible. In order to generate a maximum contraction as rapidly as possible, the neural drive from the motor control center of the brain causes instantaneous recruitment of all motor units and each motor unit is instantaneously driven at

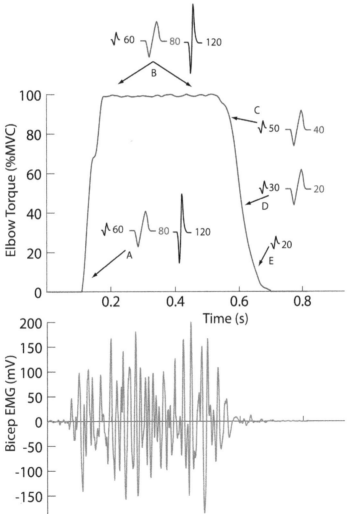

**Figure 18.10** Relationship between motor unit recruitment and firing rate changes and the resulting force–time curve and EMG amplitude for a ballistic contraction. In a ballistic contraction, the subject is instructed to generate maximum tension as rapidly as possible.

its maximal firing rate. In the ballistic contraction shown in Figure 18.10, all three motor unit types are recruited at the onset of the contraction (point A) with slow units firing at 60 Hz, intermediate units firing at 80 Hz, and fast twitch units discharging at 120 Hz. This combination of recruitment and firing rate produces a very high rate of force development and the force goes from 0 to 100 %MVC in approximately 200 ms. The subject then briefly holds the force at 100 %MVC and all three motor unit types continue to discharge at their maximal firing rate to enable the MVC force to be held (point B). Then at point C, the subject rapidly reduces the force. At point C the fast twitch units are derecruited, at point D the intermediate motor units are derecruited, and finally after point E, the slow units are derecruited.

## EFFECTS OF RATE OF FORCE DEVELOPMENT

The actual recruitment or force threshold for fast twitch motor units varies between muscles. Some muscles tend to recruit early at 60% MVC and then they will have recruited all motor units. Other muscles tend to hold very high threshold–fast twitch units until at least 80% MVC before recruiting them. As a general statement, it is safe to assume that in a ramp contraction where the force slowly goes from 0–50% MVC, the entire contraction can be performed without the use of fast twitch motor units. Figure 18.11 shows how the rate of force development can alter this general rule. In the graph on the right side of Figure 18.11, the subject generates a ramp contraction with a slow rate of force development. The subject takes approximately 2.0 seconds to go from 0% to 50% MVC. Notice that this is accomplished without the recruitment of fast twitch motor units. The curve on the left depicts the exception to this general rule that contractions at or below 50% can be performed without fast twitch motor

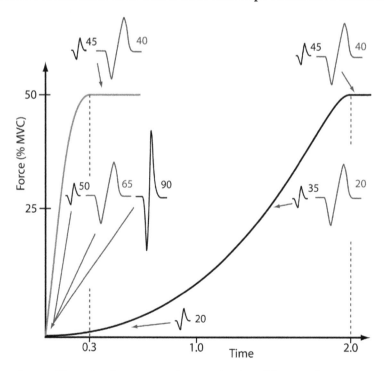

**Figure 18.11** Effects of rate of force development on the recruitment of fast twitch motor units.

units. For the curve on the left, the subject is instructed to generate a 50% MVC contraction as rapidly as possible. At the very beginning of the contraction, slow, intermediate, and fast twitch motor units are recruited with discharge rates of 50, 65, and 90 Hz, respectively. As soon as the force reaches the 50% MVC level, the fast twitch motor units are derecruited and the force level is maintained using only slow and intermediate motor units. Since fast twitch units are highly fatigable, the CNS is very protective of them. They are recruited for high force, high velocity, and eccentric contractions.

## ELECTROMECHANICAL DELAY (EMD)

A.V. Hill was the first to notice that there was a delay between the activation of a muscle and its resulting force response (Hill, 1938). When he turned on the muscle stimulator for the frog sartorius muscle, he observed a delay in the force response. In Hill's three component muscle model, the series elastic component describes this delay. Electromechanical delay (EMD) is defined as the delay between the start of EMG and the start of force. Figure 18.12 shows the electromechanical delay between the onset of bicep EMG and the start of force. The EMD in Figure 18.12 is 47 ms. The actual length of the EMD depends upon the level of activation of the muscle, the length of the muscle and the percent of fast, intermediate, and slow twitch fibers in the muscle.

### Chemical and Mechanical Events that Cause EMD

The EMG signal arises from the depolarization of the muscle membrane and it precedes the production of tension in the muscle. After $Na^+$ flows into the cell, it spreads thru the

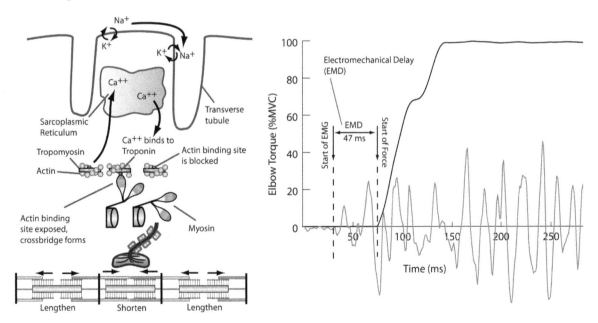

**Figure 18.12** Electromechanical delay (EMD) is the delay between the start of EMG and the start of force. The electrical and chemical events: the spread of $Na^+$ thru the cell, the release of $Ca^{++}$, the formation of an actomyosin crossbridge all occur within a few ms. The mechanical event: taking up slack by elongating titin, desmin, costamere, and the tendon account for most of the delay between the start of EMG and the start of force.

T-tubule system causing $Ca^{++}$ to be released from the SR. $Ca^{++}$ bonds with troponin causing tropomyosin to move away from the active binding site on the actin filament and therefore allowing the formation of an actomyosin crossbridge. The ATP in the myosin head is downgraded to ADP and Pi and once the Pi is released, the myosin motor uses the chemical energy obtained to do work on the actin filament. The sarcomeres directly under the endplate are the first to generate tension and this initial tension only serves to stretch or elongate the titin filaments in nearby sarcomeres. As the tension proceeds down the muscle fiber, the tendon eventually becomes tight. Once the tendon is tight, a change in force can be observed. These are the sequence of steps that explain the electromechanical delay. The chemical events occur very quickly within 1–2 ms. However, taking up slack in the series elastic component is the primary cause of EMD. In other words, tightening the titin filaments and tightening the tendon.

## CONCENTRIC AND ECCENTRIC EMG-FORCE

The EMG–force relationships presented in the ramp and ballistic contractions were limited to isometric contractions. In dynamic contractions, the relationship between EMG and force is affected by the muscle force–length relationship, force–velocity relationship, electromechanical delay, and the force generated in the actomyosin crossbridge cycle. Recall from chapter 16 that a single actomyosin crossbridge generates 10 pN of force in an isometric contraction, 4–6 pN of force in a concentric contraction, and 10–20 pN of force in an eccentric contraction. Furthermore, in a concentric contraction, the force decreases with increasing velocity of contraction and in an eccentric contraction the force increases with increasing velocity of contraction until reaching a force plateau at high eccentric velocities.

Figure 18.13 shows the resulting EMG–torque for a concentric and eccentric contraction of the biceps brachii performed on an isokinetic strength machine at 90 d/s. The isokinetic device measures the elbow torque generated at a fixed velocity (in this case, 90 d/s). The subject was instructed to generate a maximum voluntary contraction throughout the concentric and eccentric phases of the exercise. Assuming that the subject did generate an MVC, theoretically all motor units would be recruited and each motor would fire at its maximum rate. The bicep EMG has similar amplitude for both the concentric and eccentric phases, suggesting that the neural drive was constant. Since the neural input to the muscle was constant, what about the resulting output, the torque generated? Clearly, the concentric torque is less than the eccentric torque. The peak concentric torque was 61.1 N•m and the peak eccentric torque was 98.2 N•m. So how can we explain the differences in torque output? The velocity was constant and the EMG was constant, so why is the torque not constant? The answer is based upon the force generated by an individual myosin motor. Concentric activation of a motor unit results in 4–6 pN of force per actomyosin crossbridge. Eccentric activation of a motor unit results in 10–20 pN of force per actomyosin crossbridge.

We can now generalize these results to different velocities of contraction. For any fixed velocity of contraction (90 d/s for concentric, 90 d/s for eccentric), if the subject generates an MVC for the concentric and eccentric phases of the exercise, the EMG will have a constant amplitude and the resulting force or torque will be greater for the eccentric phase than the concentric phase of the exercise.

In the example above, the muscle activation was constant, which resulted in lower concentric force and higher eccentric force. Now we will example the neural input to the muscle required to generate relatively constant torque or force for both the concentric and eccentric phases of an exercise with a constant velocity of contraction. In Figure 18.14, the subject performs elbow flexion (concentric) and elbow extension (eccentric) at an approximately constant velocity while holding a 25 lb dumbbell. Notice that the peak torque is about 35 N•m for both the concentric and eccentric phase of the exercise. The biceps brachii EMG has much greater amplitude during the concentric phase than in the eccentric phase. Why did this occur? Again, the explanation is based upon the myosin motor output per crossbridge. Since less force (4–6 pN per crossbridge) is generated in a concentric contraction than in an eccentric contraction (10–20 pN per crossbridge), greater neural drive to the muscle is needed for the concentric contraction. The increased EMG amplitude shown during the concentric phase in Figure 18.14 results from some combination of additional recruitment and firing rate of motor units.

We can now generalize these results to different levels of force or torque. For any fixed velocity of contraction (90 d/s concentric, 90 d/s eccentric), if the subject generates a constant force or torque, the EMG required to generate the constant torque will have higher amplitude during the concentric phase than in the eccentric phase.

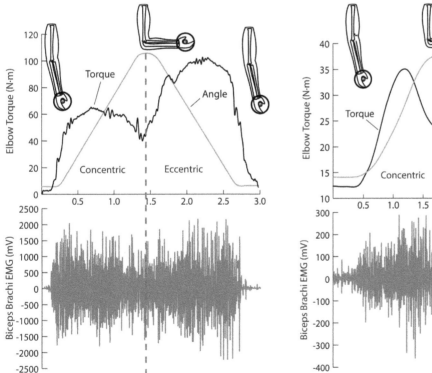

**Figure 18.13** Concentric and eccentric EMG–torque relationship. In a maximum voluntary contraction, the EMG amplitude is constant. Eccentric torque is greater than concentric torque due to higher eccentric actomyosin crossbridge force.

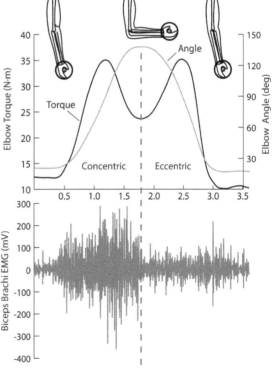

**Figure 18.14** Concentric and eccentric EMG–torque relationship. In a sub-maximal free weight curl, the concentric and eccentric peak torque is similar. Less muscle activation is required for eccentric phase due to higher eccentric actomyosin crossbridge force.

## Review Questions

1. Describe the relationship between muscle fiber types (ST, INT, FT), firing rates, twitch forces, and fatigability.
2. Draw a picture of a motor unit and show muscle fiber action potentials being propagated in each direction along the muscle fiber from the terminal branches of the nerve.
3. What is the definition of the electromyogram?
4. What factors affect the shape of the electromyogram?
5. What is the difference between a muscle fiber action potential and a motor unit action potential?
6. Give an example of how synchronization of motor units can cause amplification and cancellation of the EMG signal.
7. Draw a picture to show how stimulation or firing rate affects the force–time curve.
8. Describe the size principle.
9. Draw a picture to show EMG recruitment and firing rate changes during a ramp contraction.
10. How would motor unit recruitment and firing differ when comparing a ramp to a ballistic contraction?
11. How is recruitment threshold related to the size principle?
12. A subject generates a ramp contraction in which he/she slowly increases the force from 0–45% MVC. Describe the expected recruitment and firing rate patterns.
13. A subject generates a ballistic contraction in which he/she increases the force from 0–45% MVC as rapidly as possible. Describe the expected recruitment and firing rate patterns.
14. What is electromechanical delay? What factors are attributed to electromechanical delay? What is the primary factor that causes electromechanical delay?

## References

Henneman E. Relation between size of neurons and their susceptibility to discharge. Science. 126: 1345-1347, 1957.

Henneman E. The size-principle: A deterministic output emerges from a set of probabilistic connections. J Exp Biol. 115: 105-112, 1985.

Henneman E, Clamann HP, Gillies JD, Skinner RD. Rank order of motoneurons within a pool: Law of combination. J Neurophysiol. 37: 1338-1349, 1974.

Henneman E, Olson CB. Relations between structure and function in the design of skeletal muscles. J Neurophysiol. 28: 581-598, 1965.

Henneman E, Somjen G, Carpenter DO. Excitability and inhibitability of motoneurons of different sizes. J Neurophysiol. 28: 599-620, 1965.

Henneman E, Somjen G, Carpenter DO. Functional significance of cell size in spinal motoneurons. J Neurophysiol. 28: 560-580, 1965.

Hill AV. The heat of shortening and the dynamic constants of muscle. Proceedings of the Royal Society of London, series B - Biolgical Sciences. 117: 136-195, 1938.

## SUGGESTED SUPPLEMENTAL RESOURCES

Basmajian, J. V. and C. J. DeLuca (1985). Muscles Alive. Baltimore, Williams & Wilkins.

Enoka, R. M. (2008). Neuromechanics of Human Movement. Champaign, Human Kinetics.

Hall, S. J. (2007). Basic Biomechanics. Boston, McGraw Hill.

Hamill, J. and K. M. Knutzen (2009). Biomechanical Basis of Human Movement. Philadelphia, Lippincott Williams & Wilkins.

Kamen, G. and D. A. Gabriel (2010). Essentials of Electromyography. Champaign, Human Kinetics.

Kandel, E. R., J. H. Schwartz, et al. (2013). Principles of Neural Science. New York, McGraw-Hill.

Latash M. L. (2007). Neurophysiological Basis of Movement. Champaign, Human Kinetics.

Winter, D. A. (2009). Biomechanics and Motor Control of Human Movement. Hoboken, John Wiley & Sons, Inc.

# Chapter 19

## MOTOR RESPONSES

### LEARNING OUTCOMES

**After reading Chapter 19, you should be able to:**

- Define monosynaptic and polysynaptic motor responses.
- Describe the structure and function of the muscle spindle.
- Describe a stretch reflex.
- Describe an inverse stretch reflex.
- Describe the difference between voluntary and rapid motor responses.
- Describe the structure and function of the Golgi tendon organ.
- Describe cutaneous motor responses.
- Define a motor program.
- Define feedback and feedforward.

### REFLEX, RAPID, AND VOLUNTARY RESPONSES

Motor responses to stimuli are modulated by afferent neurons, efferent neurons, and interneurons. **Afferent neurons** transmit signals from peripheral sensors to the spinal cord. Examples of peripheral sensors include cutaneous sensors, pressure sensors, muscle spindle, and Golgi tendon organs. **Efferent neurons** transmit signals from the spinal cord to the periphery. Examples of efferent neurons include alpha motor neurons and gamma neurons. **Interneurons** are located in the spinal cord and they provide interconnections to incoming signals from the brain and periphery to the outgoing signals to the periphery. Interneurons can generate excitatory responses and inhibitory responses. A **reflex** is defined as a response to a stimulus arising from a sensory signal that cannot be modified voluntarily during its execution. Traditionally, reflexes were categorized into short latency reflexes (20–45 ms), medium latency reflexes (45–75 ms), and long latency reflexes (75–105 ms), (Lee, et al., 1975). It was originally thought that short and medium latency reflexes were invariant and that long latency reflexes could be altered by the brain stem. It is now generally accepted that sensory signals with latencies greater than 50 ms can be modified by the brain stem and the motor cortex region (Jacobs, et al., 2007; Pruszynski, et al., 2008). Returning back to our definition of the word reflex, if we define a reflex as a response to a stimulus arising from a sensory signal that cannot be modified voluntarily during its execution, there are actually very few responses that fit into this scheme. A tendon jerk (stretch reflex) from a relaxed muscle, and an eye blink are two examples that fit this criterion, as neither response can be voluntarily modified during its execution. While the word reflex is still widely used in the field of neuroscience, it now tends to be used as a generic response from a sensor (Prochazka, et al., 2000). We will classify the stretch reflex as a response 1 (R1), or short latency (20–45 ms) that cannot be modified voluntarily during its execution. We will use the term **rapid motor response** to classify motor responses from sensory signals with latencies: response 2 (R2) of 45–75 ms, and response 3 (R3) of 75 – 105 ms. Rapid motor response 2 (45–75 ms) can be modified by the brain stem and in some cases the motor cortex. Rapid motor response 3 (75–105 ms) can be modified by the motor cortex. **Voluntary motor responses** will include all motor responses with latencies of 120–180 ms.

Sensory signals arising from a muscle spindle, Golgi tendon organ and cutaneous receptors can trigger reflex, rapid motor, and voluntary responses. Figure 19.1 depicts monosynaptic (A) and polysynaptic (B) neural circuits. As shown in Figure 19.1, in monosynaptic neural circuits, there is only one synapse and in polysynaptic neural circuits, there are two or more synapses.

## PRESYNAPTIC INHIBITION AND PRESYNAPTIC FACILITATION

The efferent neural response to an incoming sensory stimulus can be altered by cortical neurons, brain stem neurons, or other sensory stimuli. These alterations in output response are accomplished using interneurons and monosynaptic connections by presynaptic inhibition and presynaptic facilitation (see Figure 19.2). In Figure 19.2, the efferent response to the incoming afferent signal from some sensory receptor can be reduced and/or completely blocked by presynaptic inhibition. **Presynaptic inhibition** occurs when an interneuron or a monosynaptic neuron releases a neurotransmitter that blocks the incoming sensory signal from stimulating the efferent neuron. **Presynaptic facilitation** can also occur via interneurons and monosynaptic projections. In presynaptic facilitation, the interneuron or the direct cortical projection releases a neurotransmitter into the synapse that will cause an increase in the efferent outflow from the incoming afferent sensory signal.

## MUSCLE SPINDLE STRUCTURE

Muscle spindles are small (2–6 mm length) fusiform shaped sensory receptors that are embedded alongside of muscle fibers and distributed throughout the muscle tissue. The structure of the muscle spindle is shown in the Figure 19.3. The muscle spindle tissue is collectively referred to as **intrafusal fibers** to differentiate it from the muscle tissue referred to as

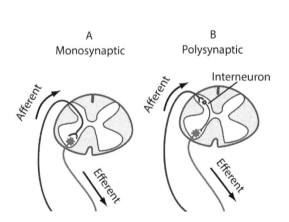

Figure 19.1 Monosynaptic and polysynaptic neural circuits.

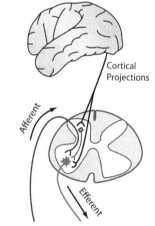

Figure 19.2 Presynaptic inhibition and facilitation arise from cortical projections.

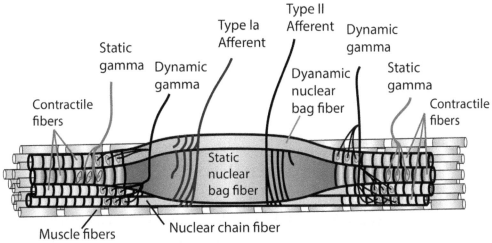

Adapted from Hulliger (1984) Rev Physiol Biochem Pharmacol, 101: 1-110.

**Figure 19.3** Muscle spindle structure. The gamma neurons innervate specialized muscle tissue on the ends of the muscle spindle which set the sensitivity of the muscle spindle.

extrafusal fibers. The muscle spindle contains three separate components: dynamic nuclear bag, static nuclear bag, and the nuclear chain fiber. Each of these three components contains a non-contractile center region that contains sensory nerve endings and a contractile region on the end of each component. The contractile regions contain specialized muscle tissue that can contract or relax to change the sensitivity of the muscle spindle. These specialized muscle tissues on the ends of each spindle component are innervated by static and dynamic gamma neurons. The dynamic gamma neurons alter the sensitivity of the muscle spindle during dynamic movements and the static gamma neurons alter the sensitivity of the muscle spindle in the static state. The muscle spindle contains two afferent fibers, the type Ia neuron (primary ending) and the type II neuron (secondary ending). The type Ia neuron, or primary ending, wraps around all three of the intrafusal fibers (dynamic nuclear bag, static nuclear bag, and nuclear chain fiber). The type II or secondary endings are spiraled around the static nuclear bag and the nuclear chain fibers only. They do not have any sensory endings that terminate on the dynamic nuclear bag. The term **fusimotor system** refers to the combination of the muscle spindle, along with its afferent (Type Ia, Type II), efferent (dynamic gamma, static gamma), and specialized muscle tissue (intrafusal fibers).

## Reflex Responses from the Muscle Spindle

Figure 19.4 shows the spinal connections for the muscle spindle. The primary (type Ia) neurons are the only sensory fibers that make direct connections with the alpha motor neurons. The type Ia neurons also make interneuron connections within the spinal cord. The secondary (type II) neurons use interneurons in the spinal cord to connect to alph motor neurons. The type Ia neurons have nerve conduction velocities of 40–90 m/s and the type II nerve fibers have nerve conduction velocities of 30–75 m/s. When a muscle spindle is stretched above its currently set sensitivity, the type Ia sends impulses to the spinal cord. These impulses are directly related to the magnitude and rate of stretch. The primary nerve endings monitor changes in length, velocity, and acceleration of the muscle and they transmit this information to the spinal cord. The static and dynamic gamma neurons are efferent neurons that innervate the specialized muscle tissue on the ends of the muscle spindle that alter the sensitivity of the spindle in static and dynamic states, respectively. Notice that the alpha motor neuron innervates the muscle fibers and the gamma neurons only innervate the specialized contractile tissue on the ends of the muscle spindle. The gamma neurons set and alter the sensitivity of the muscle spindle by causing the specialized muscle tissue on the ends of the spindle to either contract or relax.

## Stretch Reflex

Figure 19.5 illustrates the sequence of neuromuscular events that are initiated in a stretch reflex when the type Ia fibers of the muscle spindle have been stretched above their preset sensitivity level. As shown in Figure 19.5, a reflex hammer impacts the patellar tendon, which will then cause an involuntary reflex shortening of the quadriceps muscles. Typical reflex latencies for a stretch reflex via the type Ia nerve fibers are 30–50 ms. The numbers in Figure 19.5 show the time sequence of events in a stretch reflex. As shown in Figure 19.5 – [1], the patellar tendon is tapped with a reflex hammer which then causes a rapid stretch on the

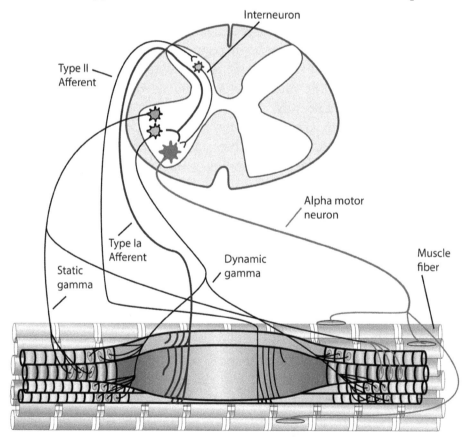

**Figure 19.4** Muscle spindle spinal connections. The Ia afferent neuron has a monosynaptic synaptic connection to the motor neuron. The II afferent neuron uses an interneuron to connect to the motor neuron.

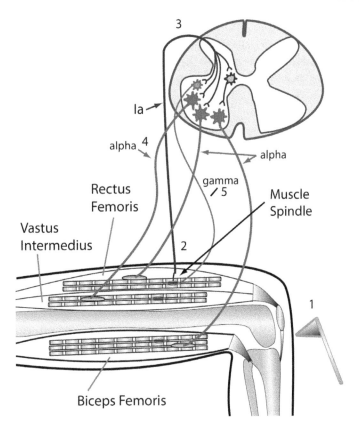

**Figure 19.5** Time sequence for neural events in a stretch reflex: 1 - patellar tendon struck with hammer, which causes a stretch of the quadriceps, 2 - Ia afferent fiber sends impulses to the spinal cord, 3 - alpha stimulation of the muscle stretched and its synergists; alpha inhibition of antagonists, 4 - alpha stimulation causes the muscle to shorten; 5 - gamma neuron causes specialized muscle tissue on the spindle to contract and reset the spindle.

rectus femoris muscle. The magnitude of stretch in the rectus femoris will determine the nature of the impulses sent to the spinal cord by the primary and secondary afferent fibers. The type Ia neurons are highly sensitive to velocity of stretch and as a result, they convey speed of movement and acceleration to the spinal cord. They are also highly sensitive to small changes in the length of the muscle. This sensitivity to small changes in length of the primary fibers is critical to the neuromuscular response to unexpected events and is particularly useful in corrective movements. In response to the stretch of the muscle caused by the impact hammer, the primary fibers will send impulses to the spinal cord (Figure 19.5 – [2]). The number and magnitude of impulses are directly related to the length and velocity of muscle stretch. The incoming Ia sensory signal from the muscle spindle arrives at the spinal cord (Figure 19.5 – [3]), where the type Ia neuron makes direct connections with the alpha motor neurons of the muscle that was stretched (rectus femoris) and its synergists (vastus intermedius, vastus lateralis, and vastus medialis). The muscle that was stretched and its synergists will receive alpha stimulation, causing them to shorten (Figure 19.5 – [4]). To facilitate this muscle shortening, the antagonists of the muscle that was stretched (biceps femoris, semitendinosus, semimembranosus) will receive alpha inhibition, causing the hamstring muscle group to relax. The reciprocal inhibition of the antagonist muscle group is accomplished via inhibitory interneurons (Geertsen, et al., 2010). The incoming sensory message from the primary ending will also stimulate the dynamic gamma neurons, as shown in Figure 19.5 – [5]. This dynamic gamma message will cause the contractile fibers on the ends of the spindle to contract to adjust the sensitivity of the spindle. The static and dynamic gamma neurons contain less myelin than do alpha motor neurons and, as a result, the nerve conduction velocity is less for the gamma neurons than the alpha motor neurons. In the next section we will examine the stretch reflex in greater detail, but first let's summarize the input-output response of the type Ia neuron in the stretch reflex. In response to a muscle stretch that causes the muscle spindle to be stretched above its current level of sensitivity, the primary (Ia) neuron sends impulses to the spinal cord which then causes the following three events to occur at the spinal level:

1. The muscle that was stretched and its synergists receive alpha stimulation. Alpha stimulation increases the activation of these muscles causing them to shorten.
2. The antagonists of the muscle that was stretched receive alpha inhibition. Alpha inhibition lowers the level of activation of the antagonists, causing them to relax. This relaxation of the antagonists makes it easier for the agonists to shorten.
3. The primary (Ia) neuron stimulates the gamma neuron. Gamma stimulation will cause the specialized muscle tissue on the ends of the muscle tissue to contract, which tightens the muscle spindle.

Figure 19.6 provides a detailed description of the stretch reflex. In step A, the muscle is stretched. If the magnitude of stretch exceeds the current sensitivity of the muscle spindle, the primary and secondary neurons will send impulses to the spinal cord. In step B, the Ia neuron transmits impulses conveying the length, velocity, and acceleration of the muscular stretch. When these neural impulses arrive at the spinal cord, they will cause three events to occur: (1) the muscle that was stretched and its synergist receive alpha stimulation, causing the muscle that was stretched and its synergist to shorten, (2) the antagonists of the muscle that was stretched receive alpha inhibition (via interneuron), which causes the antagonists to relax, and (3) the type Ia impulses generate excitatory signals in the static and dynamic gamma neurons. The excitatory signal in the dynamic gamma neurons cause the contractile fibers to contract on the dynamic nuclear bag and the nuclear chain fibers, while the excitatory signal in the static gamma neuron causes the static nuclear bag to contract. Collectively, the gamma neurons adjust the sensitivity of the muscle spindle, with the dynamic gamma neurons firing to maintain sensitivity during ongoing muscular contractions, and the static gamma neurons resetting the muscle spindle once a stable muscle length is achieved. Step C in Figure 19.6 depicts the state of the muscle spindle immediately after the muscle has shortened in response to the alpha stimulation of the muscle that was stretched and its synergists. The muscle spindle is directly connected to the muscle fibers, and as a result, when the muscle shortens, the muscle spindle also shortens. In step C, the spindle is loose or slack because of the shortening of the muscle. In addition, as soon as the spindle becomes loose, the **Ia afferent neuron stops sending impulses** to the spinal cord. Between steps C and D, the gamma stimulation causes the contractile fibers on the ends of the muscle spindle to shorten to reset the muscle spindle

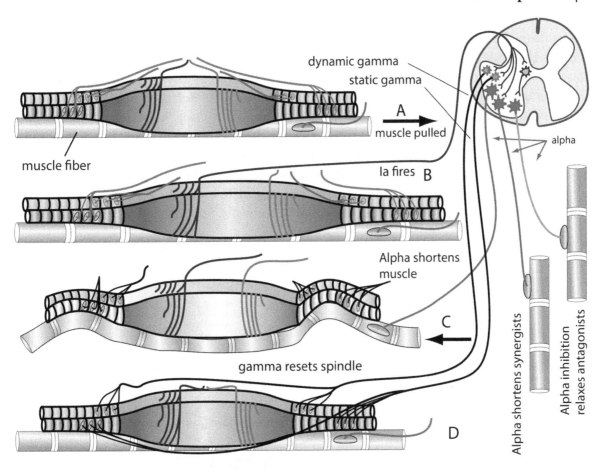

**Figure 19.6** Detailed description of the stretch reflex. A). The muscle is pulled, stretching the muscle fiber and the muscle spindle. B). Once the spindle is stretched, the Ia afferent neuron sends impulses to the spinal cord. C). At the spinal cord, the incoming Ia impulses cause alpha stimulation of the muscle stretched and its synergists, and alpha inhibition of its antagonists. C). Alpha stimulation of the muscle stretched and its synergists cause the muscle to shorten, removing the stretch on the muscle spindle, and the Ia neuron will stop sending impulses. D). The gamma neuron causes the specialized muscle tissue on the muscle spindle to contract, resetting the spindle.

so that it can accurately measure any further changes in length of the muscle. Figure 19.6 D shows the end state of the muscle spindle, after the contractile fibers have shortened the muscle spindle. It is now tight and ready to measure any future muscle length changes. Shortening of the contractile fibers on the ends of the spindle does not affect the length of the muscle tissue; it only affects the length of the muscle spindle.

## FLEXIBILITY EXERCISES AND THE INVERSE MUSCLE STRETCH REFLEX

Muscle stretching exercises are used to improve flexibility and joint range of motion. When performing stretching exercises to increase the length of a muscle, the goal is to stretch the muscle without causing a stretch reflex. It is necessary to avoid a stretch reflex, since a stretch reflex increases the activation to the muscle stretched, causing it to shorten. An inverse stretch reflex is used to stretch a muscle when performing flexibility exercises. Figure 19.7

depicts an inverse stretch reflex. The inverse stretch reflex is initiated by a cortical command in step 1. In step 1, the CNS initiates alpha and gamma inhibition. The alpha inhibitory signal reduces the muscle activation of the muscle to be stretched and its synergists, causing them to relax. The gamma inhibitory signal reduces the activation of the specialized muscle tissue on the ends of the muscle spindle, causing the spindle to relax. Relaxation of the muscle to be stretched and the muscle spindle causes the muscle and the spindle to become slack (see step 3 in Figure 19-7). The CNS command has now set the conditions to allow the muscle to be stretched. From this loosened or slack state, the muscle can now be stretched without causing a stretch reflex, as long as the rate and length of muscle stretch do not exceed the sensitivity of the muscle spindle. In step 4, the muscle is pulled to initiate a stretch. A continuous feedback loop is now established. The ongoing information from the primary and secondary afferents

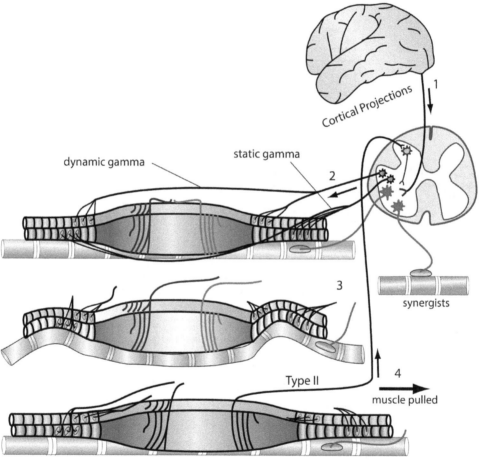

Adapted from Boyd (1980) Trends in Neurosci, 3: 258-265.

**Figure 19.7** In an inverse stretch reflex, the goal is to stretch the muscle without causing a stretch reflex. 1-2). A cortical stimulus causes alpha inhibition of the muscle to be stretched and its synergists, and inhibition of the gamma neurons to relax the muscle spindle. 2-3) Alpha inhibition relaxes the muscle, and gamma inhibition relaxes the spindle. Both are loose or slack and ready to be stretched. 4). The muscle is pulled, causing a stretch. The type II afferent signals update the brain on the current state of the muscle. A continuous feedback loop is established to enable further stretching of the muscle.

is evaluated to determine if the muscle can be elongated further. When the muscle is held at a fixed length for a brief time (10–20 seconds), it will undergo viscoelastic force relaxation, which will then enable further stretching of the muscle. If a ballistic stretch is applied at any point, it may cause a stretch reflex, which will produce the opposite result, causing the muscle to shorten rather than allowing the muscle to stretch.

## Golgi Tendon Organ Structure

Golgi tendon organs are 0.2–1 mm long sensory structures that are located at the muscle–tendon interface (see Figure 19.8). Golgi tendon organs are innervated by a type Ib afferent nerves, which have conduction velocities of 30–75 m/s, and motor response latencies of 50–80 ms. The unmyelinated nerve endings of type Ib neuron weave in and out of the collagen fibers from the tendon (Jami, 1992). A single nerve ending may be intertwined thru the tendinous bundle of 10–20 motor units. As a result of these "spaghetti–like" interconnections, a single motor unit may affect 1–6 Golgi tendon organs (Jami, et al., 1985).

Golgi tendon organs are very sensitive to changes in muscle force. Golgi tendon organs are able to respond to a twitch response from a single motor unit. Tension developed within a single motor unit causes the collagen fibers from its tendon to pinch the free–nerve endings of the Ib neuron. In response to this mechanical pinching of the nerve endings, the Ib neuron

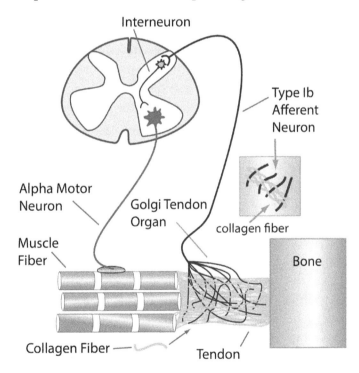

**Figure 19.8** The Golgi tendon organ monitors muscle force. The Ib afferent fibers from the Golgi tendon organ are interwoven thru the tendon's collagen fibers. Tension in the muscle fibers causes the collagen fibers to pinch the tendon organ fibers. Pinching of the fiber endings of the Golgi tendon organ causes stimuli to be transmitted to the spinal cord.

sends impulses to the spinal cord. There is a direct relationship between the level of muscle force and the number of impulses propagated up the Ib fiber (Crago, et al., 1982). Unlike type Ia afferent fibers from the muscle spindle, type Ib afferent fibers from the Golgi tendon organs do not make direct monosynaptic connections to alpha motor neurons in the spinal cord. Instead, the type Ib afferent fibers utilize polysynaptic connections to alpha motor neurons via interneurons (Jami, 1992). A further distinction of the Golgi tendon organs from muscle spindles is that the Golgi tendon organs do not contain efferent fibers and specialized muscle tissue to alter the sensitivity of the Golgi tendon organ.

## Reflex Responses from Golgi Tendon Organs

Golgi tendon organs were originally thought to have only one function, which was to protect the muscle from generating excessively high forces and potentially tearing either the muscle or tendon. It was assumed that a stimulus from the Ib neuron would cause autogenic inhibition of the muscle generating the force that elicited the stimulus. We now know that this view is far too simplistic. Simply stated, Golgi tendon organs monitor muscle force and provide force feedback to the CNS. Cortical neurons modulate the incoming Ib stimulus using presynaptic inhibition or excitation to direct the output response of the Ib stimulus. Thus, a Golgi tendon organ can cause either increased activation (excitation) or decreased activation (inhibition) of the muscle generating the tension that elicited the stimulus. The term **autogenic** refers to self-induced feedback. Therefore, in **autogenic inhibition,** the activation to a muscle experiencing high force is reduced by the Ib stimulus to lower the force in the muscle. In **autogenic excitation**, the activation to a muscle experiencing high force is increased by the Ib stimulus to raise the force in the muscle. Golgi tendon organs monitor muscle force. They are silent when there is no tension within the muscle. The Golgi tendon organ responds to both active and passive muscle tension. Active muscle tension from a single motor unit or an entire muscle results in the propagation of stimuli up the Ib neuron. Passively pulling a muscle will also elicit Ib stimuli. The number of impulses sent to the spinal cord are closely related to whole muscle force in both passive and active motions (Appenteng, et al., 1984; Crago, et al., 1982; Duysens, et al., 2000; Gregory, et al., 2002; Prochazka, et al., 2002).

Reflexes are not fixed stereotyped responses, but rather reflex response that are both task–dependent and phase–dependent. For example, in static posture, the Golgi tendon organ does cause inhibition, yet during both walking and running, the Ib input to the spinal cord causes autogenic excitation (Dietz, et al., 2000). **Autogenic excitation** is also referred to as **positive force feedback**. In positive force feedback, the stimuli from the Ib neuron cause an increase in the activation of the muscle causing the force (Prochazka, et al., 1997). Positive force feedback from the Golgi tendon organ is believed to provide stability to the lower limb during weight acceptance. Dietz and Duysens coined the term "**loading reflex**" to describe positive force feedback where Ib neural input causes alpha motor excitation during weight acceptance (Dietz & Duysens, 2000).

Figure 19.9 illustrates both positive and negative force feedback in the gastrocnemius during a plyometric jump for a skilled and unskilled jumper. The output response to the in-

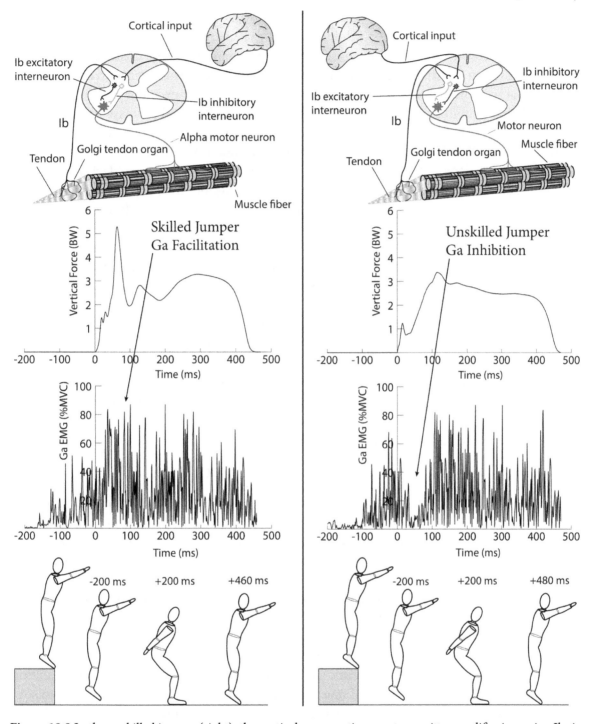

**Figure 19.9** In the unskilled jumper (right), the cortical presynaptic neurotransmitter modifies incoming Ib signal, causing autogenic inhibition (negative force feedback). The EMG signal in the unskilled jumper shows a decline in Ga activation from 50–80 ms. In the skilled jumper (left), the cortical presynaptic neurotransmitter modifies incoming Ib signal to cause autogenic facilitation (positive force feedback). The EMG signal in the skilled jumper shows an increase in Ga activation from 50–80 ms.

coming Ib neural signal is gated by the action of a cortical neuron. The cortical neuron selects the desired outcome (excitation or inhibition) using a neurotransmitter to activate the appropriate interneuron. The reflex response of the gastrocnemius for the unskilled (untrained) jumper is shown on the right. The gastrocnemius is preactivated prior to ground contact. During this preactivation period, the cortical neuron turns off the Ib excitatory interneuron and activates the Ib inhibitory interneuron. At ground contact, the rapid increase in force causes the Golgi tendon organs in the gastrocnemius to send impulses to the spinal cord. At the spinal cord level, the Ib inhibitory interneuron produces inhibition in the alpha motor neuron to the gastrocnemius. The EMG signal from the gastrocnemius shows a sharp decline in activation in the gastrocnemius from 50–80 ms. The untrained jumper uses negative force feedback (autogenic inhibition) to lower the activation of the gastrocnemius. The reflex response of the gastrocnemius for the skilled (trained) jumper is shown on the left in Figure 19.9. Once again, the gastrocnemius is pre-activated prior to ground contact. During this pre-activation period, the cortical neuron turns off the Ib inhibitory interneuron and activates the Ib excitatory interneuron. At ground contact, the rapid increase in force causes the Golgi tendon organs in the gastrocnemius to send impulses to the spinal cord. At the spinal cord level, the Ib excitatory interneuron produces excitation of the motor neuron to the gastrocnemius. The EMG signal from the gastrocnemius shows a sharp increase in activation in the gastrocnemius from 50–80 ms. The trained jumper uses positive force feedback (autogenic excitation) to increase the activation of the gastrocnemius. Notice that the vertical force for the trained jumper reaches a peak of 5.2 body weights (BW) at about 70 ms, whereas the vertical force for the untrained jumper is only at about 1.5 BW at the same time of 70 ms. This example illustrates how cortical neurons can alter the response to a stimulus.

## Muscle Stiffness Control

Muscle stiffness, which is the change in muscle force divided by the change in muscle length, is regulated by the combination of the muscle spindle, which monitors the change in length and the Golgi tendon organ, which monitors the change in muscle force. The Golgi tendon organ provides feedback of muscle force to the spinal cord and the muscle spindle provides feedback of both muscle length and velocity. For the unskilled jumper shown on the right in Figure 19.9, the muscle spindle and Golgi tendon organ were used to set the muscle stiffness during the pre-activation phase prior to landing and to respond to the change in force during the first 100 ms of landing. In this case, the subject uses sensory signals from the spindle and tendon organ to lower muscle stiffness. In comparison, the skilled jumper's response shown in Figure 19.9 illustrates how the sensory signals from the muscle spindle and Golgi tendon organ can be used to set a high level of muscle stiffness prior to ground contact and then subsequently increase the muscle stiffness during the first 100 ms after ground contact. The Ib impulses generated by the Golgi tendon organ are used for positive force feedback, which serves to increase muscle stiffness when desired (Donelan, et al., 2004; Grey, et al., 2007; Pratt, 1995; Prochazka, et al., 1997). While the Golgi tendon organ and the muscle spindle are primarily involved in regulating muscle and joint stiffness, it is safe to assume that joint receptors and cutaneous receptors will also contribute to reflex modulation of joint and muscle stiffness.

## CUTANEOUS RECEPTORS

Cutaneous reflexes play an integral role in regulating postural control and locomotion. Figure 19.10 depicts cutaneous sensors underneath the foot. Examples of cutaneous mechanoreceptors include Meissner's corpuscle, merkle disc receptor, Pacinian corpuscle, and Ruffini endings. These sensory receptors lie under the skin and they provide valuable feedback during both quiet stance and gait. During stance, cutaneous receptors on the bottom of the foot are subjected to body weight forces that vary with the sway of the center of mass. During gait, cutaneous sensors are subjected to forces that traverse the length of the foot from heel strike to toe-off. Afferent cutaneous neurons transmit sensory data to the spinal cord with a conduction velocity of 50–60 m/s, which is then used by the CNS to modulate alpha motor neurons via one or more interneurons (polysynaptic) in the spinal cord. Response latencies for cutaneous sensors vary from 50–85 ms (Aniss, et al., 1992; Burke, et al., 1991; Duysens, et al., 1993; Kanda, et al., 1983).

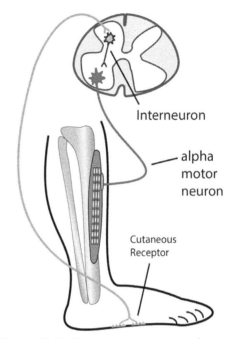

**Figure 19.10** Cutaneous receptors make connections with alpha motor neurons via one or more interneurons.

## CONVERGENCE AND MODULATION OF SENSORY SIGNALS

Figure 19.11 gives an example of how sensory signals from a variety of reflex sensors converge on a single alpha motor neuron in the spinal cord. The convergence of this rich variety of input signals can affect both the input to the motor neuron and its output response. Descending signals can originate from the several sources such as the motor cortex, brain stem, vestibular system, and visuomotor and then make either direct connections to a motor neuron or connect via interneurons. Peripheral impulses can originate from the muscle spindle, Golgi tendon organ, joint receptor, pain receptor, and cutaneous receptors. The convergence of all of these signals are affected by the context in which a motor response is evoked, the motor act being performed, and the relative time of occurrence of the stimulus event within the movement (Zehr, 2006). As shown in Figure 19.11, only the Ia neuron from the muscle spindle and the cortical neuron have a direct wiring source to the output motor neuron. All other input sources must utilize one or more interneurons to transmit their neural messages to the output motor neuron. It is this convergence of multiple input sources that leads to context and phase-sensitive motor output responses to incoming stimuli. In an isometric contraction, the neural impulses arising from the Golgi tendon organ may cause autogenic inhibition of the output motor neuron. A change of context from isometric to eccentric during landing can cause the very same Ib Golgi tendon organ impulse to elicit positive force feedback (autogenic excita-tion). Several input sources can cause this modulation of motor output responses. It could arise from the descending cortical signal, or cutaneous receptors, both of which would alter the state of the interneuron via presynaptic excitation so that incoming Ib impulse is translated from inhibition to excitation.

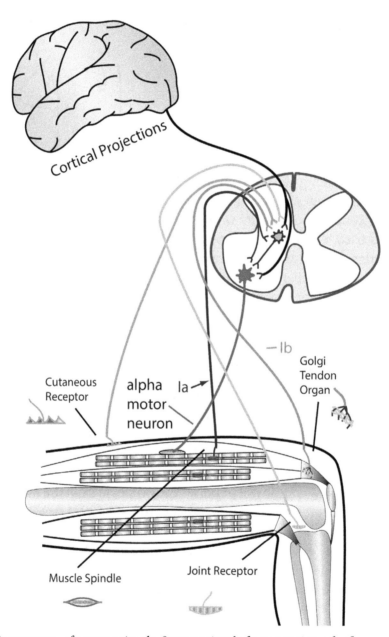

**Figure 19.11** Convergence of sensory signals. Sensory signals from a variety of reflex sensors converge on a single alpha motor neuron in the spinal cord.

## Motor Programs, Feedback, and Feedforward Control

Reflexes were once thought to be invariant and it was assumed that they were involuntary. The terms "reflex" and "voluntary" are now viewed as being on a continuum, with pure reflexes such as an eye blink or a stretch reflex on one end and pure voluntary with complete and full conscious control on the other end (Prochazka, et al., 2000). An example of a pure voluntary movement is the slow movement of your hand to reach for a stationary object, as at each phase of the motion, input from a multitude of sources is available. Furthermore, since the motion is under full conscious control, you could stop your hand at any point. In contrast, a pure reflex cannot be modified during its execution. For example, it is not possible to stop an eye blink once its execution has begun. Everything in between these two ends of the spectrum (pure reflex, conscious control) can be modulated by convergence of neural input at the spinal cord level as illustrated in Figure 19.11. A stretch reflex elicited from a relaxed muscle after a tendon jerk can also be considered as a pure reflex, as it cannot be modified during its execution. While this reflex response from a relaxed muscle cannot be altered, the response to stimuli originating from the muscle spindle Ia neuron can be overridden, attenuated, or increased when the muscle is active. At the spinal cord level, cortical neurons alter the response to incoming stimuli to produce task and phase–dependent responses. As shown in Figure 19.11, the cortical neurons can alter the potential output response either by interneurons, or by direct monosynaptic connections to the alpha motor neuron. Rapid motor responses that are observed during quiet stance can be "turned off" during movement by these cortical projections. The term **phase-dependent** is used to describe the effect where a motor response that is observed during quiet stance, is attenuated, amplified, or completely disappears during different phases (heel strike, mid stance, toe–off) of motion (Nashner, 1976; Stein, et al., 1988; Yang, et al., 1990). The term **task-dependent** is used to describe the changes in neural responses to incoming stimuli that depend upon the motor task being executed.

A **motor program** is defined as a set of muscle commands containing the desired muscle activation level and muscle timing, which are sent to the muscle before a movement begins. In fast movements, those executed in less than 50 ms, the entire movement sequence is carried out in the absence of peripheral feedback. In slower movements, those requiring more than 50 ms, various level of feedback are available to correct and modify the movement. The speed with which a movement or force is generated determines the extent to which it is able to use feedback to correct execution errors. A control system where desired outcome utilizes sensory information to compare and adjust ongoing movement or force is defined as a **feedback control**. In a feedback control movement, the movement is executed slow enough (> 50 ms) that incoming sensory signals can be used to modify or modulate the outcome of the movement (Nielsen, 2004). A control system that must generate an outcome movement or force without the availability of incorporating incoming reflex sensory information is defined as a **feedforward control** movement. In a feedforward control system, the movement is executed so fast (< 50 ms) that incoming sensory signals cannot be used to modify or modulate the outcome of the movement. Feedforward motions must be executed with pre-programmed motor sequences utilizing muscular preactivation.

### Feedforward Control

The impact force in running reaches a peak magnitude of 1–3 times your body weight in 20–40 ms. Since this event exceeds the 50 ms time window, the CNS must use feedforward control to set the desired muscle stiffness prior to ground contact. In chapter 18, we used the term **muscular preactivation** to describe the activation of a muscle prior to some know event. The level of muscular preactivation is set based upon previous experience. The CNS uses an "educated guess" to set the timing and level of preactivation, and this is known as feedforward control. Figure 19.12 shows the muscular preactivation of the gastrocnemius in barefoot running. The gastrocnemius is pre-activated approximately 80 ms prior to ground contact. During the first 50 ms of ground contact, there are no adjustments that can be made to alter the level of muscle stiffness. If the runner feels that the impact was too soft or too stiff, he or she can only make modifications to the motor program that would affect the next step. After each ground contact, the CNS makes any adjustments needed to increase or decrease joint stiffness prior to the next ground contract. In this scenario, feedback from the previous ground contact can be utilized to modify the feedforward control of future ground contact events.

### Feedback Control

The backswing in golf swing takes approximately 1.5 seconds to execute. This motion is slow enough that it can be generated using a feedback control system where sensory information can easily be used to modify the swing during its execution. For example, if during the execution of the backswing, your muscle sensors inform you that you bent your left elbow, you can respond by straightening the elbow or completely stopping the swing.

The motor response latency from a cutaneous sensor is 50–80 ms. Imagine that you are walking barefoot, very slowly across a floor in a darkened room and there is a sharp object such as a tack on the floor. If your foot is moving very slowly, there may be time for cutaneous sensors to inform you of the sharp object. Immediately upon sensing the tack, you execute an avoidance response and transfer your body weight to your opposite leg so that you can pull your foot away from the tack. This is feedback control. Now repeat the sequence but this time you are walking at a fast speed. When walking barefoot at a fast speed in a darkened room, the relatively slow response time of a cutaneous sensor would not be sufficient to protect you from injury. In this case, the force and momentum of your body weight would quickly push your foot onto the tack, causing it to puncture your skin. This example illustrates the importance of the 80–120 ms minimum time to respond to a stimulus and the difference between feedback and feedforward control.

### Combined Feedforward and Feedback Control

In some movements, the motor program uses both feedforward and feedback control to execute the movement sequence. Imagine that you are sitting at a table and there is a cup in front of you within arm's reach. Can you close your eyes and successfully reach out and pick up the cup? Give it a try, and observe the sensations experienced during the task. As soon as you close your eyes, your brain develops and subsequently executes the motor program required to reach and pick up the cup. The initial part of the program is executed using feedforward control. Your brain activates the motor units in the arm, specifying the level of force and velocity of movement required to reach in the direction of the cup. When your fingers contact the cup, you will then use feedback to complete the task to adjust your grasp so that you can grab the cup and pick it up.

## Corrective Responses

The ability to adjust motor responses based upon conditions during motion is essential to successful movement. Andrew Pruszynski describes a movement scenario where rapid motor responses are necessary to respond to both expected an unexpected events during motion (Pruszynski, et al., 2008). He describes a waitress holding a tray of drinks while moving thru a crowded restaurant. She can plan her path to avoid potential obstacles. In addition, she must also be prepared to respond to an unexpected event such as someone bumping her elbow. While traversing thru the restaurant, she observes that the table to left is empty, and the table to right is full. If her elbow is bumped, she should bias her response so that the drinks are spilled on the empty table rather than the table filled with customers. In this scenario, the waitress is using visual data to bias a rapid motor response. At the subconscious level, cortical neurons will activate the appropriate motor neurons in the event that a response is needed.

Figure 19.13 illustrates an example of a rapid motor corrective response. The subject was lifting a light weight with her eyes closed using a smooth elbow flexion motion. The subject was informed that sometime on either the upward or downward phase of the motion, that her hand would be tapped causing the arm to move in the opposite direction. The subject was instructed to resume normal upward or downward motion as soon as possible after the stimulus was applied. This is an example of a "do not interfere" instruction set. The instruction set is used by cortical neurons to bias or alter a rapid motor response. The subject in Figure 19.13 was lowering her arm from an initial elbow angle of 145° to full elbow extension of 0°. The stimulus was applied at an elbow angle of 95°. At 44 ms after the stimulus, the biceps brachii

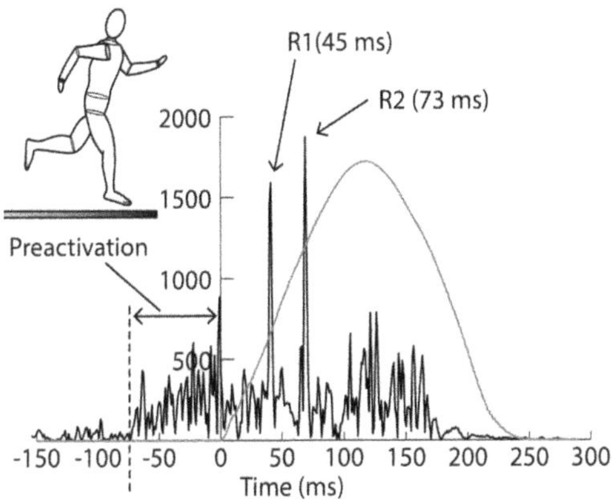


Figure 19.12 Muscle preactivation, short latency motor response (R1), and rapid latency motor response (R2) during barefoot running.

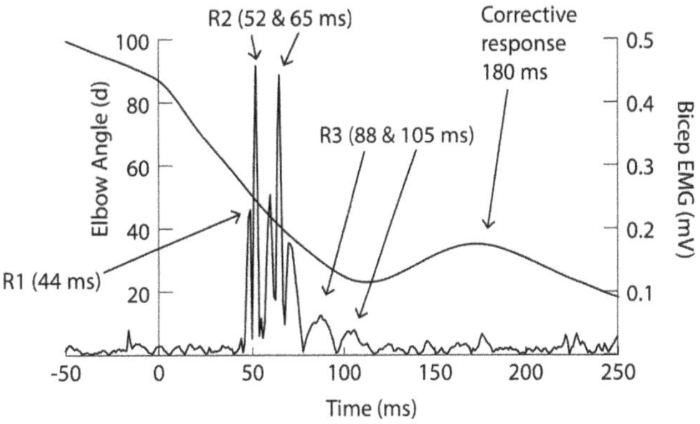


Figure 19.13 EMG corrective response. Rapid motor responses following a stimulus result in a corrective response at 180 ms.

EMG shows a stretch reflex response (R1), followed by four additional rapid motor responses. The rapid motor responses (R2) at 52 ms and 65 ms, along with the earlier R1, response attenuate the elbow angular acceleration induced by the stimulus. The cortical mediated corrective response occurring at 180 ms was produced by the rapid motor responses (R3) occurring at 88 ms and 105 ms. At 180 ms, the corrective response has been completed and the elbow motion appears to continue along the smooth path as if nothing happened. The subject successfully executed the "do not interfere" instruction set.

## SUMMARY

This chapter characterizes the structure and role of sensory receptors on subsequent motor neuron output. The basic functioning of the muscle spindle and the Golgi tendon organ are presented along with task–dependent and phase–dependent motor responses.

## REVIEW QUESTIONS

1. Describe a stretch reflex.
2. List the afferent and efferent fibers on a muscle spindle and a Golgi tendon organ.
3. What is the difference between a stretch reflex and an inverse stretch reflex.
4. Give an example of a motion that uses feedback and feedforward control.

## REFERENCES

Aniss AM, Gandevia SC, Burke D. Reflex responses in active muscle elicited by stimulation of low-threshold afferents from the human foot. Journal of Neurophysiology. 67: 1375-1384, 1992.

Appenteng K, Prochazka A. Tendon organ firing during active muscle lengthening in awake, normally behaving cats. Journal of Physiology. 353: 81-92, 1984.

Burke D, Dickson HG, Skuse NF. Task-dependent changes in the responses to low-threshold cutaneous afferent volleys in the human lower limb. Journal of Physiology. 432: 445-458, 1991.

Crago PE, Houk JC, Rymer WZ. Sampling of total muscle force by tendon organs. J Neurophysiol. 47: 1069-1083, 1982.

Dietz V, Duysens J. Significance of load receptor input during locomotion: A review. Gait and Posture. 11: 102-110, 2000.

Donelan JM, Pearson KG. Contribution of force feedback to ankle extensor activity in decerebrate walking cats. J Neurophysiol. 92: 2093-2104, 2004.

Duysens J, Clarac F, Cruse H. Load-regulating mechanisms in gait and posture: Comparative aspects. Physiological Reviews. 80: 83-133, 2000.

Duysens J, Tax AA, Trippel M, Dietz V. Increased amplitude of cutaneous reflexes during human running as compared to standing. Brain Research. 613: 230-238, 1993.

Geertsen SS, Stecina K, Meehan CF, Nielsen JB, Hultborn H. Reciprocal Ia inhibition contributes to motoneuronal hyperpolarisation during the inactive phase of locomotion and scratching in the cat. The Journal of Physiology. 589: 119-134, 2010.

Gregory JE, Brockett CL, Morgan DL, Whitehead NP, Proske U. Effect of eccentric muscle contractions on golgi tendon organ responses to passive and active tension in the cat. J Physiol. 538: 209-218, 2002.

Grey MJ, Nielsen JB, Mazzaro N, Sinkjær T. Positive force feedback in human walking. The Journal of Physiology. 581: 99-105, 2007.

Jacobs JV, Horak FB. Cortical control of postural responses. J Neural Transm. 114: 1339-1348, 2007.

Jami L. Golgi tendon organs in mammalian skeletal muscle: Functional properties and central actions. Physiological Reviews. 72: 623-666, 1992.

Jami L, Petit J, Proske U, Zytnicki D. Responses of tendon organs to unfused contractions of single motor units. J Neurophysiol. 53: 32-42, 1985.

Kanda K, Sato H. Reflex responses of human thigh muscles to non-noxious sural stimulation during stepping. Brain Research. 288: 378-380, 1983.

Lee RG, Tatton WG. Motor responses to sudden limb displacements in primates with specific cns lesions and in human patients with motor system disorders. Canadian Journal of Neurological Sciences. 2: 285-293, 1975.

Nashner LM. Adapting reflexes controlling the human posture. Experimental Brain Research. 26: 59-72, 1976.

Nielsen JB. Sensorimotor integration at spinal level as a basis for muscle coordination during voluntary movement in humans. J Appl Physiol. 96: 1961-1967, 2004.

Pratt CA. Evidence of positive force feedback among hindlimb extensors in the intact standing cat. Journal of Neurophysiology. 73: 2578-2583, 1995.

Prochazka A, Clarac F, Loeb GE, Rothwell JC, Wolpaw JR. What do reflex and voluntary mean? Modern views on an ancient debate. Exp Brain Res. 130: 417-432, 2000.

Prochazka A, Gillard D, Bennett DJ. Positive force feedback control of muscles. Journal of Neurophysiology. 77: 3226-3236, 1997.

Prochazka A, Gritsenko V, Yakovenko S. Sensory control of locomotion: Reflexes versus higher-level control. In: Gandevia SG, Proske U, Stuart DG, editors. Sensori-motor control. New York: Kluwer Academic/Plenum. p. 1-13, 2002.

Pruszynski JA, Kurtzer I, Scott SH. Rapid motor responses are appropriately tuned to the metrics of a visuospatial task. Journal of Neurophysiology. 100: 224-238, 2008.

Stein RB, Capaday C. The modulation of human reflexes during functional motor tasks. Trends in Neuroscience. 11: 328-332, 1988.

Yang JF, Stein RB. Phase-dependent reflex reversal in human leg muscles during walking. Journal of Neurophysiology. 63: 1109-1117, 1990.

Zehr EP. Training-induced adaptive plasticity in human somatosensory reflex pathways. Journal of Applied Physiology. 101: 1783-1794, 2006.

## Suggested Supplemental Resources

Basmajian, J. V. and C. J. DeLuca (1985). Muscles Alive. Baltimore, Williams & Wilkins.

Enoka, R. M. (2008). Neuromechanics of Human Movement. Champaign, Human Kinetics.

Hall, S. J. (2007). Basic Biomechanics. Boston, McGraw Hill.

Hamill, J. and K. M. Knutzen (2009). Biomechanical Basis of Human Movement. Philadelphia, Lippincott Williams & Wilkins.

Kamen, G. and D. A. Gabriel (2010). Essentials of Electromyography. Champaign, Human Kinetics.

Kandel, E. R., J. H. Schwartz, et al. (2013). Principles of Neural Science. New York, McGraw-Hill.

Latash M. L. (2007). Neurophysiological Basis of Movement. Champaign, Human Kinetics.

Shumway-Cook, A. and M. H. Woollacott (2006). Motor Control: Translating Research into Clinical Practice. Philadelphia, Lippincott Williams & Wilkins.

Winter, D. A. (2009). Biomechanics and Motor Control of Human Movement. Hoboken, John Wiley & Sons, Inc.

# Chapter 20

## BALANCE AND POSTURAL CONTROL

### STUDENT LEARNING OUTCOMES

**After reading Chapter 20, you should be able to:**

- Define postural control.
- Define sensory organizational theory.
- Define the role of somatosensory, vestibular, and vision in postural control.
- Describe the sensory organization test.
- Describe the conundrum between healthy bone and joint issues in the elderly.

### SENSORY INFORMATION AND POSTURAL CONTROL

To maintain upright standing posture in state of equilibrium, the central nervous system (CNS) must keep the center of mass within the boundaries of the base of support. Sounds simple enough, but unfortunately for the CNS, there is no system in the body that is exclusively designed to measure the location of the center of mass with respect to the base of support. The CNS regulates balance by monitoring data from three primary sources: somatosensory, vision, and vestibular. The CNS then evaluates the data from each source and effectively performs a cross-check to scan for errors in the data. If one or more of the systems (somatosensory, vision, and vestibular) indicates that the center of mass is no longer in balance, the CNS must select an appropriate motor response to reposition the center of mass within the boundaries of the base of support. Figure 20.1 depicts the three systems of the body that the CNS uses to modulate static and dynamic balance. The somatosensory system includes the muscle spindle, Golgi tendon organ, cutaneous receptors, and joint receptors. In response to a postural challenge, the somatosensory system initiates automated or triggered reflex responses which can be overridden by direct cortical projections or interneurons at the spinal cord level. Somatosensory information is transmitted to the CNS via ascending tracts in the spinal cord. The vestibular system consists of specialized sensory organs located in the inner ear that measure both linear and angular accelerations of the head. Linear and angular accelerations of the head, detected by the vestibular sensory organs, are used by the CNS to determine the direction of motion and the body's orientation in space. Accelerations of the head can elicit rapid motor responses via vestibulospinal pathways that project onto motor neurons. Finally, the CNS uses vision to determine body orientation, direction, velocity of motion, and environmental factors that affect static and dynamic equilibrium. The CNS uses the visuomotor system to respond to visually detected postural challenges.

### SENSORY ORGANIZATIONAL THEORY

According to the sensory organizational theory, the brain gathers somatosensory, vestibular, and vision data to determine the body's orientation and state equilibrium. The brain then evaluates available data to resolve any conflicting information and subsequently select an appropriate response to maintain balance in desired position and equilibrium (Nashner, 1989). The brain must weigh visual, vestibular, and somatosensory signals to determine the accuracy of each and then adaptively select an appropriate response to any postural challenge. Somatosensory and visual signals are more sensitive to postural challenges, than are vestibular signals and, as a result the CNS predominantly relies upon both vision and somatosensory data to modulate postural control. Vision and somatosensory sensors are more likely to provide inaccurate or misleading information. When presented with inaccurate or conflicting sensory data, the CNS resolves conflicting information to determine which signals to ignore and which sensory signals should be used to develop an appropriate motor program to respond to the postural challenge. The selected motor program is then executed to return the center of mass over the base of support.

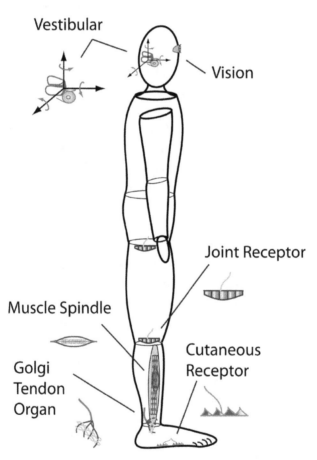

**Figure 20.1** In theory, maintaining balance is simple: just keep your center of mass positioned over the base of support. Unfortunately, there is no single system in the body that provides the CNS with the location of the center of mass relative to the base of support. Instead the CNS, must use sensory information flowing in from visual, somatosensory, and vestibular sensors, verify the accuracy of the incoming signals, and then initiate an appropriate neuromuscular response to modulate balance.

Drivers often experience an example of inaccurate visual information. A driver is stopped at a red light and the car beside the driver moves backward. Instantaneously the driver's peripheral vision informs the driver that his car has moved forward and the driver reflexively applies the brakes. After the driver has applied the brakes, he realizes that his car was not moving. It was the car beside him that was moving and the vision system provided misleading information. To err on the side of safety, the CNS initiated the motor program to depress the brakes while resolving the conflicting information in the background. After the brakes were applied, the CNS retroactively evaluated both somatosensory and vestibular signals which led to the conclusion that the driver's car was not moving. The vestibular signals indicated that the head was not accelerating and somatosensory receptors indicated that the pressure sensations from the seat and back also verified that the body was not accelerating.

## REFLEX LATENCY AND POSTURAL CONTROL SENSORY SIGNALS

Somatosensory signals are faster and more abundant than are vestibular and visual sensory data. Reflex latencies from muscle spindles are typically 30–50 ms and latencies from Golgi tendon organs, cutaneous sensors, and joint sensors are typically 50–80 ms (Diener, et al., 1988; Horak, et al., 1989; Moore, et al., 1988; Nashner, 1976; Nashner, 1977; Nashner, 1979).

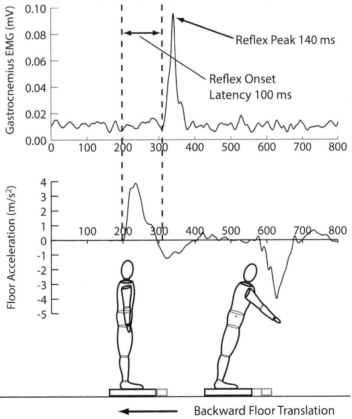

**Figure 20.2** Rapid motor response elicited during the motor control test. A backward horizontal floor translation causes ankle plantar flexion. A long latency rapid motor response of the gastrocnemius begins 100 ms after the stimulus is applied.

The fastest visuomotor reflex response is approximately 100 ms (Nashner, et al., 1978). Accelerations of the head have been shown to elicit vestibular induced muscular reflex responses with a latency of 40–84 ms (Horak, et al., 1994; Horstmann, et al., 1988).

Figure 20.2 shows an example of a rapid motor response that is elicited in the gastrocnemius following a backward horizontal floor translation. The test begins with the subject balanced in quiet stance. The floor is translated backward, which then causes the subject's center of mass to be translated in the forward direction. The forward translation of the subject center of mass causes ankle plantar flexion, which stretches the gastrocnemius. The EMG signal of the gastrocnemius shows a rapid motor response onset of 100 ms, with the peak response occurring 140 ms after the stimulus is applied.

## SENSORY ORGANIZATION TEST (SOT)

The sensory organization test (SOT) was developed by Lewis Nashner. The SOT evaluates postural control using a sequence of six testing conditions to test somatosensory and visual contributions and indirectly tests vestibular contributions to balance. The sensory organization test is performed using a NeuroCom Smart Equitest© system, which was developed by Nashner to be used in physical therapy clinics and research laboratories for balance evaluation and testing.

The NeuroCom Smart Equitest system is composed of a force plate that is mounted on a moving system that can translate the floor forward and backward and rotate the floor in a toes–up or toes–down direction. The floor rotation mechanism is used to provide inaccurate somatosensory information during the balance test. Figure 20.3 illustrates the six conditions of the sensory organization test that is used to evaluate and or train postural control.

### Inaccurate Somatosensory Information

The force plates output the location of the subject's center of pressure, which is then used to determine the position of the subject's center of mass relative to the base of support. To understand how the Smart Equitest can provide intentionally inaccurate somatosensory information, try this simple test. Stand up with your hands at your side. From a balanced position, slowly lean forward and pay attention to the sensation of pressure under your feet. As you move forward, your cutaneous sensors inform your CNS that your center of pressure has moved toward your toes. Now here is how the Smart Equitest can provide inaccurate somatosensory information: when you lean forward you should feel the center of pressure moving in the direction of your toes. However, the Smart Equitest rotates the floor in the toe-downward direction exactly equal to the amount of forward shift of your center of mass. As a result of the floor rotating in unison with your forward lean, the sensation of pressure from your cutaneous sensors is unchanged and your CNS thinks everything is fine, when in fact you are about to fall forward. The system works in the opposite direction when you lean backward, it rotates your toes up so that your cutaneous sensors again mislead you into believing that everything is fine, when in fact, you are about to fall backward.

## Inaccurate Visual Information

The Smart Equitest provides misleading visual data to the CNS by rotating the walls of the system as the subject leans forward or backward. The system is composed of three walls (front, right, and left side) similar to a telephone booth. The walls are referred to as a moveable visual surround surface. When the subject leans forward, the system rotates the surround sur-

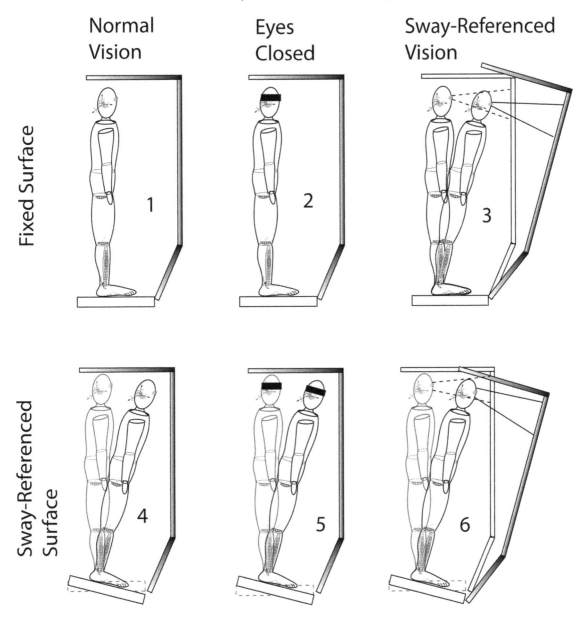

**Figure 20.3** Conditions in the NeuroCom Sensory Organization Test (SOT). Condition 1: eyes open, stable floor and walls. Condition 2: eyes closed, stable floor. Condition 3: sway–referenced surround (walls) provides misleading visual information. Condition 4: sway–referenced floor provides misleading somatosensory information. Condition 5: no vision, misleading somatosensory information; subject must use vestibular. Condition 6: inaccurate visual and somatosensory information with sway–referenced surround and floor.

face forward exactly equal to the magnitude of forward lean. The visual information provided to the CNS indicates that everything is fine when in fact, the subject is about to fall forward. When the subject leans backwards, the surround surface also rotates background again providing misleading visual information to the CNS.

### Sensory Organization Test Conditions

In the SOT test, postural control is evaluated by assessing an individual's ability to use accurate sensory information and suppress inaccurate sensory information (see Figure 20.3). The SOT test evaluates an individual's balance using the following six conditions:

1. Eyes open, fixed surface, and fixed visual surround.
2. Eyes closed, fixed surface.
3. Eyes open, fixed surface, sway-referenced visual surround.
4. Eyes open, sway-referenced surface, fixed visual surround.
5. Eyes closed, sway referenced surface.
6. Eyes open, sway-referenced surface, sway-referenced visual surround.

## DYNAMIC BALANCE AND POSTURAL CONTROL

The sensory organization test described above is often referred to as a measure of dynamic balance. It is dynamic in the sense that the floor rotates and the surround system rotates to test the subject's response to a postural challenge while standing. One of the limitations of the SOT is that it is dynamic postural challenge that is applied in a static balanced position. Since reflex responses are task–dependent and phase–dependent, the reflex responses observed during static stance may not, and in most instances, do not operate in the same manner when the individual is walking, running, climbing up or down stairs.

To maintain balance during static stance, the CNS must keep the center of mass within the boundaries of the base of support. Dynamic balance during walking is a completely different task for the CNS to control. In walking, the CNS must collect somatosensory, visual, vestibular, and momentum data. Walking is composed of phases of single support and double support. During the double support phase, the trailing leg transfers momentum to the leading leg. In addition to transferring momentum from the trail leg to the lead leg, the CNS must also transfer the momentum in the medial–lateral direction to shift the weight from the right foot to the left foot. Figure 20.4 shows an elderly subject walking with his eyes closed. A comparison of the path of the subject's center of mass with eyes open and eyes closed is shown in Figure 20.5. The medial–lateral deviations of the subject's center of mass between each right heel contact (RHC) are greater with eyes closed than with eyes open. The increased medial–lateral deviations during eyes closed walking relate to impaired dynamic balance. While humans don't routinely walk around with their eyes closed, this does suggest that dynamic balance may be impaired in elderly subjects when walking in low light conditions.

## Postural Control and Reducing Fall Risk in the Elderly

Muscular strength and vision decline with aging. This decline in strength and vision along with other factors such as medication contributes to an increased risk of falls in the elderly. As pointed out in chapter 15, bone mineral density peaks somewhere between age 30 to 40 and thereafter bone mineral density declines each year. In the elderly cortical bone thickness is decreased, there are fewer trabecular plates and fewer trabecular rods. Collectively all of these changes in bone result in frail tissue that can easily be broken. In addition the osteogenic effect of exercise declines with age. Finally, many elderly individuals have some level of joint deterioration which makes high intensity exercise painful. Despite these challenges age and health appropriate exercises such as fast walking, vibration platforms and geriatric exercise can minimize the likelihood of falls or reduce the rate of decline in muscular strength and bone structure and orientation. Training programs for the elderly have been shown to improve strength and balance and reduce the incidence of falls in the elderly (Alexander, et al., 2005; Allum, et al., 1998; Carpenter, et al., 2006; Horak, 2006; Liu-Ambrose, et al., 2006; Maki, et al., 2008; Marigold, et al., 2005; Silsupadol, et al., 2006; Wolf, et al., 2003; Woollacott, 2000).

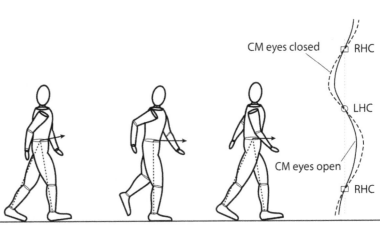

**Figure 20.4** Example of dynamic postural control: walking with eyes closed.

**Figure 20.5** Dynamic postural control. During gait, momentum is transferred from the trial leg to the lead leg and from the right side to the left side. The graph shows the medial–lateral deviations of the center of mass when walking with eyes open and eyes closed. RHC is right heel contact and LHC is left heel contact.

## Definition of Terms

Limits of Stability – the outermost region that the center of mass can be placed while still maintaining static or dynamic balance.

Postural Control – controlling the body's position to maintain desired orientation and equilibrium while standing (static equilibrium) and moving (equilibrium).

Sensory Organizational Theory – the CNS gathers somatosensory, vestibular, and vision data to determine the body's orientation and equilibrium state, resolving any conflicting information, and then determines appropriate response to maintain balance in desired position and equilibrium.

Somatosensory – body sensory signals arising from the muscle spindle, Golgi tendon organ, cutaneous receptors, and joint receptors.

Vestibular – receptor organs located in the inner ear that detect linear and angular accelerations of the head.

## Review Questions

1. Describe sensory information that the brain uses to organize a response necessary to maintain balance.
2. An individual is standing on a moving platform with his eyes closed. Describe the role of vestibular and somatosensory information available to the brain to organize postural control responses.

## References

Alexander NB, Ashton-Miller JA, Giordani B, Guire K, Schultz AB. Age differences in timed accurate stepping with increasing cognitive and visual demand: A walking trail making test. J Gerontol A Biol Sci Med Sci. 60: 1558-1562, 2005.

Allum JHJ, Bloem BR, Carpenter MG, Hulliger M, Hadders-Algra M. Proprioceptive control of posture: A review of new concepts. Gait and Posture. 8: 214-242, 1998.

Carpenter MG, Adkin AL, Brawley LR, Frank JS. Postural, physiological and psychological reactions to challenging balance: Does age make a difference? Age and Ageing. 35: 298-303, 2006.

Diener HC, Horak FB, Nashner LM. Influence of stimulus parameters on human postural responses. Journal of Neurophysiology. 59: 1888-1905, 1988.

Horak FB. Postural orientation and equilibrium: What do we need to know about neural control of balance to prevent falls? Age and Ageing. 35: ii7-11, 2006.

Horak FB, Diener HC, Nashner LM. Influence of central set on human postural responses. J Neurophysiol. 62: 841-853, 1989.

Horak FB, Shupert CL, Dietz V, Horstmann G. Vestibular and somatosensory contributions to responses to head and body displacements in stance. Experimental Brain Research. 100: 93-106, 1994.

Horstmann GA, Dietz V. The contribution of vestibular input to the stabilization of human posture: A new experimental approach. Neuroscience Letters. 95: 179-184, 1988.

Liu-Ambrose T, Khan KM, Donaldson MG, Eng JJ, Lord SR, McKay HA. Falls-related self-efficacy is independently associated with balance and mobility in older women with low bone mass. J Gerontol A Biol Sci Med Sci. 61: 832-838, 2006.

Maki BE, Cheng KCC, Mansfield A, Scovil CY, Perry SD, Peters AL, et al. Preventing falls in older adults: New interventions to promote more effective change-in-support balance reactions. Journal of Electromyography and Kinesiology. 18: 243-254, 2008.

Marigold DS, Eng JJ, Dawson AS, Ingilis JT, Harris JE, Gylfadóttir S. Exercise leads to faster postural reflexes, improved balance and mobility, and fewer falls in older persons with chronic stroke. Journal of the American Geriatrics Society. 53: 416-423, 2005.

Moore SP, Rushmer DS, Windus SL, Nashner LM. Human automatic postural responses: Responses to horizontal perturbations of stance in multiple directions. Experimental Brain Research. 73: 648-658, 1988.

Nashner L. Sensory, neuromuscular, and biomechanical contributions to human balance. In: Duncan PW, editor. Proceedings of the apta forum. Alexandria, VA: APTA. p. 5-12, 1989.

Nashner L, Berthoz A. Visual contribution to rapid motor responses during postural control. Brain Research. 150: 403-407, 1978.

Nashner LM. Adapting reflexes controlling the human posture. Exp Brain Res. 26: 59-72, 1976.

Nashner LM. Fixed patterns of rapid postural responses among leg muscles during stance. Exp Brain Res. 30: 13-24, 1977.

Nashner LM. Organization and programming of motor activity during posture control. Prog Brain Res. 50: 177-184, 1979.

Silsupadol P, Siu KC, Shumway-Cook A, Woollacott MH. Training of balance under single- and dual-task conditions in older adults with balance impairment. Phys Ther. 86: 269-281, 2006.

Wolf SL, Sattin RW, Kutner M, O'Grady M, Greenspan AI, Gregor RJ. Intense tai chi exercise training and fall occurrences in older, transitionally frail adults: A randomized, controlled trial. Journal of the American Geriatrics Society. 51: 1693-1701, 2003.

Woollacott MH. Systems contributing to balance disorders in older adults. The Journals of Gerontology Series A: Biological Sciences and Medical Sciences. 55: M424-M428, 2000.

## SUGGESTED SUPPLEMENTAL RESOURCES

Basmajian, J. V. and C. J. DeLuca (1985). Muscles Alive. Baltimore, Williams & Wilkins.

Enoka, R. M. (2008). Neuromechanics of Human Movement. Champaign, Human Kinetics.

Hall, S. J. (2007). Basic Biomechanics. Boston, McGraw Hill.

Hamill, J. and K. M. Knutzen (2009). Biomechanical Basis of Human Movement. Philadelphia, Lippincott Williams & Wilkins.

Kamen, G. and D. A. Gabriel (2010). Essentials of Electromyography. Champaign, Human Kinetics.

Kandel, E. R., J. H. Schwartz, et al. (2013). Principles of Neural Science. New York, McGraw-Hill.

Latash M. L. (2007). Neurophysiological Basis of Movement. Champaign, Human Kinetics.

Shumway-Cook, A. and M. H. Woollacott (2006). Motor Control: Translating Research into

Clinical Practice. Philadelphia, Lippincott Williams & Wilkins.

Winter, D. A. (2009). Biomechanics and Motor Control of Human Movement. Hoboken, John Wiley & Sons, Inc.

# APPENDIX A ANSWERS TO REVIEW QUESTIONS

## CHAPTER 2 ANSWERS

1. Define a scalar and give an example of a scalar. A scalar is a quantity that only has magnitude. Mass and energy are examples of a scalars.
2. Define a vector and give an example of a vector. A vector is a quantity has both direction and magnitude. Examples of vector quantities include position, velocity, acceleration, force, torque, and momentum.
3. A force of 100 N is applied at an angle of 60 degrees. Draw the vector nents.

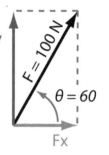

4. A soccer ball is accelerated with a horizontal component Ax = 7 m/s² and Ay = 14 m/s², draw the x, y components and the acceleration vector direction and magnitude.

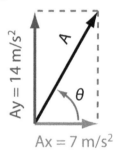

5. Compute the x and y components of each vector. Draw the vector and the x, y components.
   a. 200 N at 30 deg

$$x = r(cos\theta)$$
$$Fx = 200(cos\,30)$$
$$Fx = 173.21N$$

$$y = r(sin\theta)$$
$$Fy = 200(sin\,30)$$
$$Fy = 100.00N$$

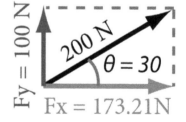

b.  4 m/s at 220 deg  [x = −3.06 m/s, y = −2.57 m/s]

$$x = r(cos\theta)$$
$$Px = 4(cos\,220)$$
$$Px = -3.06m/s$$

$$y = r(sin\theta)$$
$$Py = 4(sin\,220)$$
$$Py = -2.57m/s$$

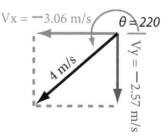

6. Compute the magnitude and direction of each vector. If the vector is not in the first quadrant convert the angle so that it is defined counterclockwise from the right horizontal.
   a. Fx = 200 N, Fy = 1400 N

$$r = \sqrt{x^2 + y^2}$$
$$r = \sqrt{200^2 + 1400^2}$$
$$r = 1414.21 N$$

$$\theta = tan^{-1}\left(\frac{y}{x}\right)$$
$$\theta = tan^{-1}\left(\frac{1400}{200}\right)$$
$$\theta = 81.87\, deg$$

   b. Ax = −2 m/s², Ay = −4 m/s²

$$\theta = tan^{-1}\left(\frac{y}{x}\right)$$
$$\theta = tan^{-1}\left(\frac{-4}{-2}\right)$$

$$r = \sqrt{x^2 + y^2}$$
$$r = \sqrt{-2^2 + -4^2}$$
$$r = 4.47 m/s^2$$

$$\theta = 63.43\, deg$$
$$\theta = 63.43 + 180\, deg$$
$$\theta = 243.43\, deg$$

   c. Vx = −3 m/s, Vy = 6 m/s

$$\theta = tan^{-1}\left(\frac{y}{x}\right)$$

$$r = \sqrt{x^2 + y^2}$$
$$r = \sqrt{-3^2 + 6^2}$$
$$r = 6.71 m/s$$

$$\theta = tan^{-1}\left(\frac{6}{-3}\right)$$
$$\theta = -63.43\, deg$$
$$\theta = -63.43 + 180\, deg$$
$$\theta = 116.57\, deg$$

7. Add the vector pairs shown below. Give a graphical picture of the vector addition and compute the magnitude and direction of the resultant vector. If the resultant vector is not in the first quadrant convert the angle so that it is defined counterclockwise from the right horizontal.
   a. Vector A = 300 N at 50 degrees, Vector B = 100 N at 25 degrees. [magnitude = 392.91 N, direction = 43.83°]
   b. Vector A = 10 m at 310 degrees, Vector B = 3 m at 140 degrees. [magnitude = 7.06 m, direction = 305.77°]

7a. Vector A = 300 N at 50 degrees, Vector B = 100 N at 25 degrees.

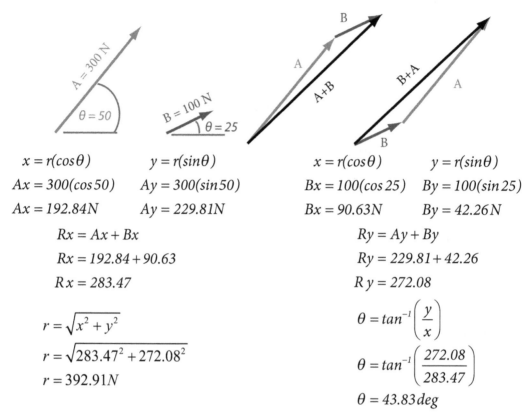

$x = r(cos\theta)$    $y = r(sin\theta)$

$Ax = 300(cos50)$    $Ay = 300(sin50)$

$Ax = 192.84N$    $Ay = 229.81N$

$Rx = Ax + Bx$

$Rx = 192.84 + 90.63$

$Rx = 283.47$

$r = \sqrt{x^2 + y^2}$

$r = \sqrt{283.47^2 + 272.08^2}$

$r = 392.91N$

$x = r(cos\theta)$    $y = r(sin\theta)$

$Bx = 100(cos25)$    $By = 100(sin25)$

$Bx = 90.63N$    $By = 42.26N$

$Ry = Ay + By$

$Ry = 229.81 + 42.26$

$Ry = 272.08$

$\theta = tan^{-1}\left(\frac{y}{x}\right)$

$\theta = tan^{-1}\left(\frac{272.08}{283.47}\right)$

$\theta = 43.83deg$

7b. Vector A = 10 m at 310 degrees, Vector B = 3 m at 140 degrees.

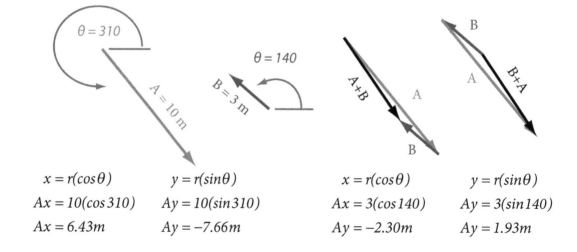

$x = r(cos\theta)$    $y = r(sin\theta)$

$Ax = 10(cos310)$    $Ay = 10(sin310)$

$Ax = 6.43m$    $Ay = -7.66m$

$x = r(cos\theta)$    $y = r(sin\theta)$

$Ax = 3(cos140)$    $Ay = 3(sin140)$

$Ay = -2.30m$    $Ay = 1.93m$

$$Rx = Ax + Bx$$
$$Rx = 192.84 + 90.63$$
$$Rx = 283.47$$

$$Ry = Ay + By$$
$$Ry = -7.66 + 1.93$$
$$Ry = -5.73$$

$$r = \sqrt{x^2 + y^2}$$
$$r = \sqrt{4.13^2 + -5.73^2}$$
$$r = 7.06m$$

$$\theta = tan^{-1}\left(\frac{y}{x}\right)$$
$$\theta = tan^{-1}\left(\frac{-5.73}{4.13}\right)$$
$$\theta = -54.23\,deg$$
$$\theta = -54.23 + 360$$
$$\theta = 305.77\,deg$$

## CHAPTER 3 ANSWERS

Before answering these questions remember that you are looking at a velocity curve which represents the change in SLOPE for position and the slope of the velocity curve represents the acceleration.

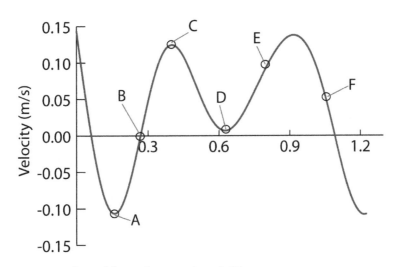

**Figure 3.14** Velocity-time graph used for review questions 1-15.

1.  The position at point A is? The velocity at point A is negative which means that the position must have a downward slope. Since point A is a negative peak it is the steepest downward region of position over the entire red interval.
2.  The acceleration at point A is? The slope at point A is horizontal therefore the acceleration at point A is 0.
3.  The position at point B is? The velocity is zero at point B, therefore position must be at a horizontal slope, since the velocity is increasing at point B, the position for point B must be at a local maximum.

4. The acceleration at point B is? The slope of velocity at point B is steep upward the acceleration must be positive.

5. The position at point C is? The velocity at point C is positive which means that the position must have a upward slope. Since point C is a positive peak it is the steepest upward region of position over the entire blue interval.

6. The acceleration at point C is? The velocity at point C has a horizontal slope, therefore acceleration at point C is 0.

7. The position at point D is? The velocity at point D is positive, so the position must be increasing. Since the velocity at point D is positive, but CLOSE to zero the position will increase very slowly, it will be a shallow upward spot for position.

8. The acceleration at point D is? The velocity at point D has a horizontal slope, therefore acceleration at point D is 0.

9. The position at point E is? The velocity at point E is positive, so the position must be increasing.

10. The acceleration at point E is? The slope of the velocity at point E is upward so acceleration must be positive.

11. The position at point F is? The velocity at point F is positive so the slope of position must be upward.

12. The acceleration at point F is? The slope of velocity at point F is downward, therefore the acceleration will be negative.

13. How will the negative velocity area affect position? Since the velocity is negative over the interval the position must always decrease. At the end of the negative interval the position will be lower than at the start of the negative interval.

14. How will the positive velocity area affect position? Since the velocity is positive over the interval the position must always increase over the interval. At the end of the positive interval the position will be higher than at the start of the positive interval.

15. When you look at the entire velocity curve, is there more positive or negative areas, and how will this affect the change in position over the time interval from start t=0 to finish t= 1.2? There is more positive area than negative area, therefore, the position will be higher at t = 1.2, than at t=0.

The graph on the next page shows the complete position, velocity and acceleration curves for Figure 3.14.

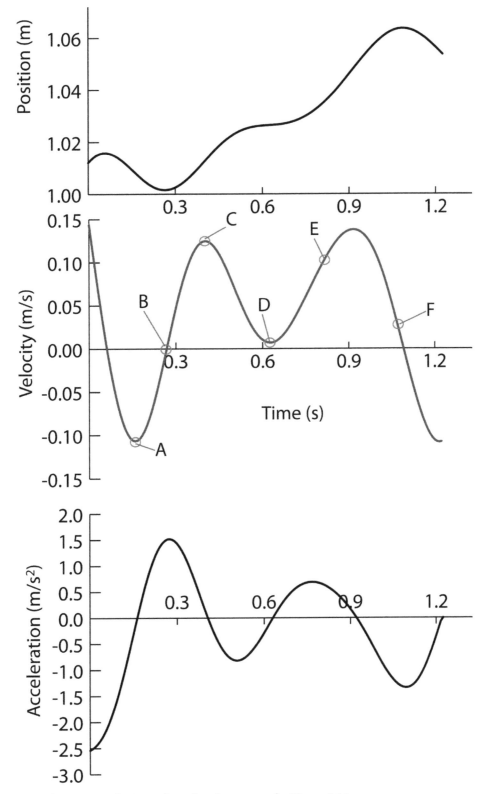

**Figure 3.14** Actual position, velocity and acceleration curves for Figure 3.14.

## CHAPTER 5 ANSWERS

## REVIEW CONCEPT ANSWERS

1. Draw a picture of a projectile that takes off and lands at the same height. Show the horizontal and vertical velocity at the following time points: takeoff, midway on the ascent, at the peak, midway on the descent, at landing. Describe the relationship between time up, time down and the total time in the air.

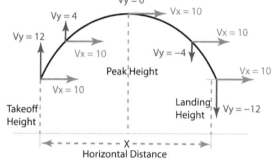

When an object takes off and lands at the same height the time from takeoff to peak is equal to the time from peak to landing. At equal heights the vertical velocity is equal and opposite.

2. Draw a picture of a projectile that takes off at a height of 14 m lands at a height of 0 m. Show the horizontal and vertical velocity at the following time points: takeoff, midway on the ascent, at the peak, midway on the descent, at landing. Describe the relationship between time up, time down and the total time in the air.

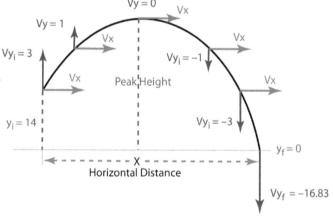

The time from takeoff to peak will be less than the time from peak to landing. When the object gets back down to the height of 14 m its vertical velocity will be equal in magnitude to the take-off velocity. Since the object has to fall the additional 14 m it will have a greater vertical velocity at landing than takeoff.

3. A golf ball is hit at an initial height of 0 m and it takes off with a vertical velocity of 18 m/s and a horizontal velocity of 27 m/s, what is the horizontal velocity of the ball at the instant that it strikes the ground?

If we ignore the effects of air resistance than the horizontal acceleration for the golf ball in the air is 0 m/s². Since the horizontal acceleration is 0 then horizontal velocity must be constant. Therefore the golf ball will have a horizontal velocity of 27 m/s when it strikes the ground.

## REVIEW PROBLEMS

1. A ball is thrown with a vertical velocity ($Vy_i$) of 13 m/s, horizontal velocity (Vx) of 15 m/s and an initial height ($y_i$) of 1.2 m.

   a. Find the height of the ball after 2 seconds.

   $$y_f = y_i + Vy_i(t) + \frac{1}{2}at^2$$

   $$y_f = 1.2 + 13(2) + \frac{1}{2}(-9.8)(2)^2$$

   $$y_f = 1.2 + 13(2) + \frac{1}{2}(-9.8)(2)^2$$

   $$y_f = 1.2 + 26 - 19.6$$

   $$y_f = 7.68m$$

   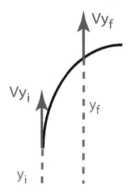

   b. Find the vertical velocity at 2 seconds

   $$Vy_f = Vy_i + at$$

   $$Vy_f = 13 - 9.8(2)$$

   $$Vy_f = -6.6m/s$$

   c. Find the horizontal distance covered in 2 seconds

   $$x = Vx(t)$$

   $$x = 15(2)$$

   $$x = 30m$$

2. A triple jumper reaches a peak height ($y_f$) of 2.1 m, if her takeoff height ($y_i$) was 1.1 m what was her takeoff vertical velocity.

   $$(Vy_f)^2 = (Vy_i)^2 + 2a(y_f - y_i)$$

   $$(0)^2 = (Vy_i)^2 + 2(-9.8)(2.1 - 1.1)$$

   $$0 = (Vy_i)^2 - 19.6$$

   $$19.6 = (Vy_i)^2$$

   $$\sqrt{19.6} = \sqrt{(Vy_i)^2}$$

   $$Vy_i = 4.43m/s$$

   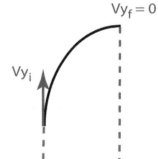

3. A football in the air has a height ($y_i$) of 27 m and a vertical velocity ($Vy_i$) of −2 m/s, what is the vertical velocity of the ball when it hits the ground (final height of 0 m).

$$(Vy_f)^2 = (Vy_i)^2 + 2a(y_f - y_i)$$

$$(Vy_f)^2 = (-2)^2 + 2(-9.8)(0 - 27)$$

$$(Vy_f)^2 = 4 + 529.2$$

$$\sqrt{(Vy_f)^2} = \sqrt{533.2}$$

$$Vy_f = 23.09m/s$$

$$Vy_f = -23.09m/s$$

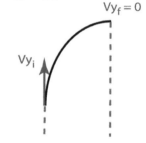

4. A baseball leaves the bat with a vertical velocity ($Vy_i$) of 34 m/s and an initial height ($y_i$) of .8 m, how many seconds will it take for the ball to reach peak height ($y_r$)?

$$Vy_f = Vy_i + at$$

$$0 = 34 - 9.8(t)$$

$$9.8(t) = 34$$

$$t = \frac{34}{9.8}$$

$$t = 3.47s$$

5. A soccer ball lands with a vertical velocity ($Vy_f$) of −13 m/s and a final height ($y_f$) of 0 m. What was the peak height of the ball?

$$(Vy_f)^2 = (Vy_i)^2 + 2a(y_f - y_i)$$

$$(-13)^2 = (0)^2 + 2(-9.8)(0 - y_i)$$

$$169 = 19.6 y_i$$

$$\frac{169}{19.6} = y_i$$

$$y_i = 8.62m$$

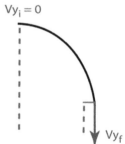

6. It's a windy day at Fenway Park in Boston. A Red Sox player hits a baseball with an initial height $y_i$ of 0.9 m with a horizontal velocity ($Vx$) of 23 m/s and an initial vertical velocity ($Vy_i$) of 25 m/s, what horizontal distance did the ball travel if it lands with a final height ($y_f$) 0 m? [x = 118.22 m]

This is a three part problem: first find $Vy_f$ at landing, then find the time in the air, finally find the horizontal distance.

**$Vy_f$ at landing**

$$(Vy_f)^2 = (Vy_i)^2 + 2a(y_f - y_i)$$
$$(Vy_f)^2 = (25)^2 + 2(-9.8)(0 - .9)$$
$$(Vy_f)^2 = 625 + 17.64$$
$$\sqrt{(Vy_f)^2} = \sqrt{642.64}$$
$$Vy_f = 25.35m/s$$
$$Vy_f = -25.35m/s$$

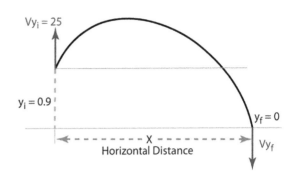

**Now find the time in the air.**

$$Vy_f = Vy_i + at$$
$$-25.35 = 25 - 9.8(t)$$
$$9.8(t) = 50.35$$
$$t = \frac{50.35}{9.8}$$
$$t = 5.14s$$

**Now find horizontal distance.**

$$x = Vx(t)$$
$$x = 23(5.14)$$
$$x = 118.22m$$

## CHAPTER 6 ANSWERS

1.  A softball player throws the ball with an initial angular velocity ($\omega_i$) of 2.2 r/s, after 0.6 seconds the players final angular velocity ($\omega_f$) is 3.3 r/s, compute the angular acceleration ($\alpha$). Draw a picture of the angular velocities and the angular acceleration.

$$\alpha = \frac{\omega_f - \omega_i}{\Delta t}$$
$$\alpha = \frac{3.3 - 2.2}{0.6}$$
$$\alpha = 1.83r/s^2$$

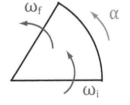

2.  Calculate the linear velocity ($v$) obtained if a golfer swings a golf club with a radius of 1.2 m using a angular velocity ($\omega$) of 330 d/s.

$$radians = \frac{deg\,rees}{57.29}$$

$$\omega = \frac{330d\,/\,s}{57.29d\,/\,r}$$

$$\omega = 5.76r\,/\,s$$

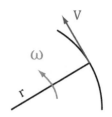

$$v = \omega r$$

$$v = (5.76r\,/\,s)(1.2m)$$

$$v = 6.91m\,/\,s$$

3.  Calculate the linear velocity ($v$) obtained if a golfer swings a golf club with a radius of 0.7 m using a angular velocity ($\omega$) of 330 d/s.

$$v = \omega r$$

$$v = (5.76r\,/\,s)(0.7m)$$

$$v = 4.03m\,/\,s$$

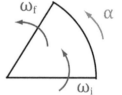

4.  A javelin throwers forearm has an initial angular velocity ($\omega_i$) of 6.3 r/s and a radius of 0.7 m.  Compute the angular acceleration ($\alpha$) necessary to increase his angular velocity ($\omega_f$) to 6.8 r/s in a time of 0.3 s.

$$\alpha = \frac{\omega_f - \omega_i}{\Delta t}$$

$$\alpha = \frac{6.8 - 6.3}{0.3}$$

$$\alpha = 1.67r\,/\,s^2$$

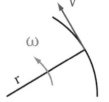

5.  A discus thrower rotates his arm with an initial linear velocity ($V_i$) of 7.7 m/s and a radius of 1.2 m about the shoulder axis. After 0.6 seconds the club has a final linear velocity ($V_f$) of 9.3 r/s.  Find the tangential acceleration ($a_T$).  Draw a picture showing the velocity components and the tangential angular velocity.

$$a_T = \frac{V_f - V_i}{t}$$

$$a_T = \frac{9.3 - 7.7}{0.6}$$

$$a_T = 2.67m\,/\,s^2$$

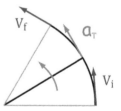

6.  When throwing a 12 kg hammer, a track athlete accelerates ($\alpha$) the hammer −5.5 r/s² for a time of 0.4 sec with a radius ($r$) of 1.3 m. Compute the tangential acceleration ($a_T$). Draw a picture of the angular and tangential acceleration vectors.

$$a_T = \alpha r$$

$$a_T = (-5.5r/s^2)(1.3m)$$

$$a_T = -7.15m/s^2$$

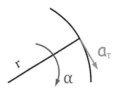

7. A bowler rotates his arm with a radius of 1.3 m and a linear velocity (V) of 6.9 m/s. Compute the centripetal acceleration ($a_c$) and centripetal force ($F_c$) necessary to keep the 10 kg bowling ball in the circle. Draw a picture of the linear velocity, centripetal acceleration and the centripetal force.

$$a_C = \frac{V^2}{r}$$

$$a_C = \frac{(6.9m/s)^2}{1.3m}$$

$$a_C = 36.62m/s^2$$

$$F_C = m\frac{V^2}{r}$$

$$F_C = 10kg\frac{(6.9m/s)^2}{1.3m}$$

$$F_C = 366.23N$$

8. Compute the centripetal acceleration ($a_c$) and centripetal force ($F_c$) for a 80 kg runner going around a track turn with a radius (r) of 12 m and an angular velocity ($\omega$) of 1.9 r/s. Draw a picture of the linear velocity, centripetal acceleration and the centripetal force.

$$a_C = r\omega^2$$

$$a_C = (12m)(1.9r/s)^2$$

$$a_C = 43.32m/s^2$$

$$F_C = mr\omega^2$$

$$F_C = (10kg)(12m)(1.9r/s)^2$$

$$F_C = 3465.6N$$

9. A runner contacts the ground with an initial lower leg angle ($\theta_i$) of 0.4 r, after a time of 0.3 seconds the final lower leg angle is ($\theta_f$) is −.25 r. Compute the angular velocity ($\omega$).

$$\omega = \frac{\theta_f - \theta_i}{\Delta t}$$

$$\omega = \frac{-.25 - 0.4}{0.3}$$

$$\omega = -2.17r/s$$

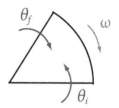

## Chapter 7 Answers

1. What is the difference between an applied force and a reaction force? Which force accelerates your center of mass the applied force or the reaction force?

   When doing a vertical jump you push down on the ground. This downward force is the applied force, it accelerates the in the direction of the applied force. The reaction force is in the opposite direction of the applied force with equal magnitude. Your center of mass is accelerated in the direction of the applied force.

2. State Newton's three laws.

   **Law of Inertia**: a body at rest stays at rest, a body in motion stays in motion in a straight line unless acted upon by a net force.

   **Law of Acceleration**: the acceleration that a body experiences is directly proportional to the net force, inversely proportional to its mass and occurs in the direction of the force. The law of acceleration is described by the equation $F = m\,a$.

   **Law of Reaction**: for every action there is an opposite and equal reaction.

3. Draw a free body diagram of a sprinter positioned in the blocks (both feet and both hands are in contact with the ground.

   Begin by drawing the force of gravity at the center of mass. Then include a horizontal and vertical reaction force for any points of contact with the ground or object. Assume the direction for each Rx and Ry force is positive.

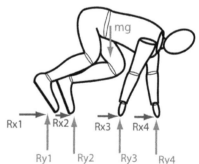

4. Draw a free body diagram of a high jumper in the air.

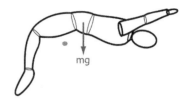

The force of gravity is the only force acting on the jumper in the air (assuming that we are ignoring the effects of air resistance).

5. Define inertia.

Inertia is the natural tendency of an object to remain at rest or in motion at a constant speed in a straight line.

6. If the net reaction forces Rx and Ry acting on a runner are in the positive x direction and positive y direction what will be the directions for $a_x$ and $a_y$ of the runner's center of mass?

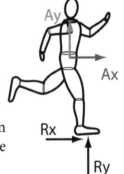

The direction of $a_x$ and $a_y$ of the runner's center of mass are always in the direction of the net force. Therefore ax and ay will both be in the positive direction.

7. An 85 kg runner experiences a horizontal acceleration ax of −5.9 m/s² and a vertical acceleration ay of 18.6 m/s². Complete the FBD and solve for the horizontal (Rx) and vertical (Ry) reaction forces that caused these accelerations.

$$\Sigma Fx = ma_x$$
$$Rx = ma_x$$
$$Rx = (85kg)(-5.9m/s^2)$$
$$Rx = -501.5N$$
$$\Sigma Fy = ma_y$$
$$Ry + mg = ma_y$$
$$Ry + (85kg)(-9.8) = (85kg)(18.6m/s^2)$$
$$Ry = 2414N$$

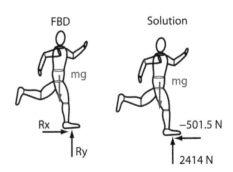

8. A 70 kg triple jumper experiences a horizontal acceleration ax of 7.4 m/s² and a vertical acceleration ay of 14.3 m/s². Complete the FBD and solve for the horizontal (Rx) and vertical (Ry) reaction forces that caused these accelerations.

$$\Sigma Fx = ma_x$$
$$Rx = ma_x$$
$$Rx = (70kg)(7.4m/s^2)$$
$$Rx = 518N$$

$$\Sigma Fy = ma_y$$
$$Ry + mg = ma_y$$
$$Ry + (70kg)(-9.8) = (70kg)(14.3m/s^2)$$
$$Ry = 1687N$$

FBD          Solution

Rx

Ry          518 N          1687 N

9. A 50 kg gymnast experiences a horizontal reaction force (Rx) of −520 N and a vertical reaction force (Ry) of 1210 N. Complete the free-body diagram and determine the horizontal ($a_x$) and vertical ($a_y$) accelerations of the gymnast's center of mass caused by these reaction forces.

$$\Sigma Fx = ma_x$$
$$Rx = ma_x$$
$$-520 = (50kg)(a_x)$$
$$a_x = -10.4m/s^2$$

$$\Sigma Fy = ma_y$$
$$Ry + mg = ma_y$$
$$1210 + (50kg)(-9.8) = (50kg)(a_y)$$
$$a_y = 14.4m/s^2$$

FBD          Solution

Rx          −520

Ry          1210

## CHAPTER 8 ANSWERS

1. Define linear momentum. What are the units for linear momentum? How is the direction of the vector determined?

   Linear momentum ($p$) is mass × velocity. The units for linear momentum are kg·m/s. The vector for linear momentum points in the same direction as the velocity. The equation for linear momentum is

$$p = mv$$

2. Define impulse. What are the units for impulse? How is the direction of the impulse vec-

tor determined? Give two methods for computing impulse.

Impulse is defined as the product of the average force and the time interval $\Delta t$ during which the force acts. Impulse is also defined as the area underneath the force – time curve. Impulse (J) is a $\bar{F}$ vector quantity, it has the same direction as the average force and the units for impulse are N·s. The following two formulas define the impulse.

$$J = \bar{F}\Delta t$$

Where J is impulse, is the average force and $\Delta t$ is the time which the force acts. The effect of multiplying the average force by time is to compute the area, or impulse in N·s under the average force curve. Since impulse is the area under the force time curve the following equation also gives the impulse.

$$J = \int_{t_0}^{t_1} F \, dt$$

3. State the principle of conservation of linear momentum.

The total linear momentum of a system of objects is constant if the net external force acting on the system is zero. The total linear momentum is defined by the following equation.

$$m_A u_A + m_B u_B = m_A v_A + m_B v_B$$

4. What is the difference between an elastic collision and an inelastic collision?

**Elastic collision**: in an elastic collision the two objects collide and then rebound. **Inelastic collision**: in an inelastic collision the two objects collide and stick together.

5. Compute the linear momentum for a 12 kg ball that has a velocity of 4.7 m/s.

$$p = mv$$
$$p = (12kg)(4.7m/s)$$
$$p = 56.4kg \cdot m/s$$

6. A 0.4 kg ($m_A$) billiard ball with an initial velocity ($u_A$) of 2.3 m/s collides with a 0.2 kg ($m_B$) billiard ball with an initial velocity ($u_B$) of −2.9 m/s in a perfectly elastic collision. Determine the velocity of each billiard ball after the collision.

$$v_A = \frac{m_B u_B}{m_A} \qquad\qquad v_B = \frac{m_A u_A}{m_B}$$

$$v_A = \frac{(0.2kg)(-2.9m/s)}{0.4kg} \qquad\qquad v_B = \frac{(0.4kg)(2.3m/s)}{0.2kg}$$

$$v_A = -1.45kg \cdot m/s \qquad\qquad v_B = 4.60kg \cdot m/s$$

7. A 1.6 kg ($m_A$) piece of soft clay with an initial velocity ($u_A$) of 4.2 m/s collides with another piece of soft clay with a mass ($m_B$) of 0.9 kg and an initial velocity ($u_B$) of −2.3 m/s in a perfectly inelastic collision. Determine the velocity (v) of the pieces of clay after the collision.

$$m_A u_A + m_B u_B = (m_A + m_B)v$$
$$(1.6kg)(4.2m/s) + (0.9kg)(-2.3m/s) = (1.6kg + 0.9kg)v$$
$$v = 1.86kg \cdot m/s$$

8.  State the impulse – momentum equation.

$$\text{Force} \times \text{time} = \text{change in linear momentum}$$
$$\bar{F}(\Delta t) = mV_f - mV_i$$
$$\int_{t_0}^{t_1} F \, dt = mV_f - mV_i$$

The integral of force with respect to time is equal to the change in linear momentum. In other words, the area underneath the force – time curve (which is impulse) is equal to the change in linear momentum.

9.  An average force of 85 N is applied for 0.15 seconds to a 1.2 kg ball. Prior to applying the force the ball has an initial velocity ($V_i$) of 0 m/s, determine the impulse applied, the change in momentum and the final velocity ($V_f$) of the ball.

Impulse is force × time

$$J = \bar{F}\Delta t$$
$$J = (85N)(.15s)$$
$$J = 12.76N \cdot s$$

Impulse-Momentum equation

$$\bar{F}(\Delta t) = mV_f - mV_i$$
$$(85N)(.15s) = (1.2kg)(V_f) - (1.2kg)(0m/s)$$
$$V_f = 10.63m/s$$

## CHAPTER 9 ANSWERS

1.  State both the average force and the integral versions of the impulse – momentum equation. For each equation explain how you could compute impulse using both the right and left hand side of the equation.

Impulse can be computed either by measuring the force and the time of force application, or by measuring the change in velocity before and after the force is applied and computing the change in momentum.

$$\text{Average Force} \times \text{Time} = \text{Change in Linear Momentum}$$

The left hand side of the equation above computes impulse by multiplying average force times time. The right hand side of the equation above computes impulse by computing the change in linear momentum.

$$\bar{F}(\Delta t) = mV_f - mV_i$$

The Area under the Force – Time Curve = Change in Linear Momentum

The left hand side of the equation above computes impulse by computing the area underneath the force – time curve. The right hand side of the equation above computes impulse by computing the change in linear momentum.

$$\int_{t_0}^{t_1} F \, dt = mV_f - mV_i$$

2. Describe the relationship between braking and propulsion impulses in running and the horizontal velocity of the runners center of mass.

A braking force will cause the runner's horizontal velocity to decrease. A propulsion force will cause a runner's horizontal velocity to increase.

3. Describe the steps necessary and the equations needed to compute horizontal acceleration and horizontal velocity from a horizontal force – time curve in running.

To compute horizontal acceleration from force – time data use Newton's law of acceleration and divide the force by mass.

$$\Sigma Fx = ma_x$$

$$a_x = \frac{\Sigma Fx}{m}$$

To compute horizontal velocity from force – time data use either the average force or integral version of the horizontal impulse momentum equation and solve for horizontal velocity.

$$\bar{R}x(\Delta t) = mVx_f - mVx_i$$

$$\int_{t_0}^{t_1} Fx \, dt = mVx_f - mVx_i$$

4. Describe the steps necessary and the equations needed to compute vertical acceleration, vertical velocity from a vertical force – time curve in a vertical jump.

To compute vertical acceleration from force – time data use Newton's law of acceleration and divide the force by mass.

$$\Sigma Fy = ma_y$$

$$a_y = \frac{\Sigma Fy}{m}$$

To compute vertical velocity from force – time data use either the average force or integral version of the vertical impulse momentum equation and solve for vertical velocity.

$$(\bar{R}y + mg)(\Delta t) = mVy_f - mVy_i$$

$$\int_{t_0}^{t_1} Fy \, dt = mVy_f - mVy_i$$

5. Use the horizontal force curve in Figure 9.21 to answer questions a-f.

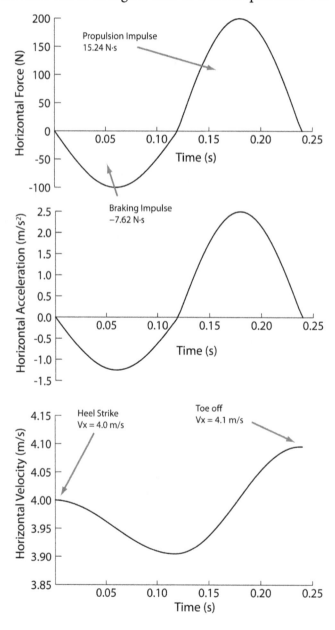

a. Draw an xy graph of what you think the horizontal acceleration would look like.

The horizontal acceleration graph will have the same shape as the horizontal force graph. Dividing the force by mass gives acceleration in m/s².

b. Draw an xy graph of what you think the horizontal velocity curve would look like. As- sume that the initial horizontal velocity is 4 m/s.

The horizontal velocity must decrease when the horizontal force is negative and the in- crease when the horizontal force is positive.

c. Which impulse is greater braking or propulsion?

The negative force – time area is less than the positive force – time area, therefore the brak- ing impulse is less than the propulsion impulse.

d. At toe off will the runner's horizontal velocity be greater or less than the initial hori- zontal velocity of 4 m/s at heel strike?

Since propulsion is greater than braking the runner's horizontal velocity at toe off will be greater than the runner's horizontal velocity at heel strike.

## CHAPTER 10 ANSWERS

1. Define work and power.

   Work is force × distance. Work can be computed using the following formulas.
   $$W = F(y_f - y_i)$$
   $$W = \frac{1}{2}m(Vy_f)^2 - \frac{1}{2}m(Vy_i)^2 + mgh_{y_f - y_i}$$
   $$W = \int_{t_0}^{t_1} F \times \Delta y$$

   Power is the rate of doing.   Power is a scalar and the units for power are watts (W).  Power can be computed using either of the formulas below.
   $$P = \frac{W}{t}$$
   $$Power = Force \times Velocity$$

2. How is kinetic, potential and total mechanical energy calculated?

   Kinetic energy is the energy a body possesses due to its motion.  Kinetic energy is calcu- lated as follows.

$$KE = \frac{1}{2}mVy^2$$

Potential energy is the energy a body possesses due to its position. Potential energy is calculated as follows.

$$PE = mgh$$

3. What is the difference between the work – energy principle and the law of conservation of energy? Under what conditions does each apply?

   The work – energy principle states that the work done is equal to the change in energy. The work – energy principle only applies when you are in contact with the ground or some fixed object where you can apply a force. The reaction to the force you apply does work on you and changes your energy.

   The law of conservation of energy states that when gravity is the only force operating on a system the total energy of the system is constant (conserved). The law of conservation of energy only applies when an object or performer is in the air and gravity is the only force, (this assumes that air resistance is negligible).

4. Define rotational work and describe how rotational work is calculated.

   Rotational work done by a constant torque rotating an object some angular distance is given by:

$$W_R = \tau(\theta_f - \theta_i)$$

   There are three different methods for computing the rotational work for a non-constant torque. (1) Determine the average torque and then multiply the average torque by the angular displacement. (2) Use integration to determine the area underneath the torque – angle graph. (3) Compute power from torque and angular velocity and then integrate the power curve with respect to time.

5. A 75 kg skate boarder takes off with a vertical velocity (Vy) of 2.9 m/s and a height (yi) of 1.2 m. Find the peak height using energy.

   This is a two part problem first find kinetic, potential and total energy at takeoff. Then use the total energy to find the peak height.

$$KE = \frac{1}{2}mVy^2$$

$$KE = \frac{1}{2}(75kg)(2.9m/s)^2$$

$$KE = 315.38J$$

$$PE = mgh$$

$$PE = (75kg)(9.8)(1.2m)$$

$$PE = 882J$$

Now compute the total energy, then use law of conservation of energy to find peak height.

$$E = KE + PE$$

$$E = 315.38 + 882$$

$$E = 1197.38J$$

$$h = \frac{E}{mg}$$

$$h = \frac{1197.38J}{(75kg)(9.8)}$$

$$h = 1.63m$$

6. A 50 kg dancer lands with an initial vertical velocity ($Vy_i$) of −2.8 m/s and an initial height of 1.22 m. The dancer applies a force against the ground to reduce her momentum. After 0.5 s she has a final vertical velocity ($Vy_f$) 0.0 m/s and a final height ($y_f$) of 0.88 m. Compute the work done and average power.

$$W = \frac{1}{2}m(Vy_f)^2 - \frac{1}{2}m(Vy_i)^2 + mgh_{y_f-y_i}$$

$$W = \frac{1}{2}(50)(0.0)^2 - \frac{1}{2}(50)(-2.8)^2 + (50)(9.8)(0.88-1.22)$$

$$W = -362.6J$$

$$P = \frac{W}{t}$$

$$P = \frac{-362.6J}{0.5s}$$

$$P = 724.4W$$

## CHAPTER 11 ANSWERS

1. Define torque and moment of inertia.

   Torque ($\tau$) is a force that causes or tends to cause rotation about some point or axis of rotation. Torque is a vector with its direction defined by the right hand rule and units of N·m. Moment of inertia (I) represents an objects resistance to angular change. Moment of inertia is a scalar and it is a function of mass times squared distance about some axis of rotation.

2. State Newton's laws of angular motion.

   **I – Law of Inertia** – A body at rest stays at rest, a rotating body stays in rotation unless acted upon by a net torque.
   **II – Law of Acceleration** – The angular acceleration experienced by a body is directly proportional to the net torque and inversely proportional the moment of inertia and it occurs in the direction of the net torque. The law of acceleration is explained by the following equation. $\Sigma\tau = I\alpha$
   **III – Law of Reaction** – For every torque there is an opposite and equal reaction torque.

3. List the steps in computing torque.

(1) Take the magnitude of the force (ignore + or –). (2) Find the line of action of the force about the point of torque calculation (or axis of rotation). (3) Determine the lever arm, which is the perpendicular distance from the line of action of the force to the point of torque calculation (or axis of rotation). (4) Determine the direction of torque the force will cause or tend to cause about the point of torque calculation (axis of rotation). (5) $\tau = |F| \times l$ Multiply the force magnitude by the lever arm and give the product a negative value if for a CW or a positive value for a CCW rotation caused by the force about the point.

4. Using Figure 11.14 below list the lever arm for each force about points A, B, and C.

**Point A**: the lever arm is 0.6 m for –50 N force, 0.0 m for –60 N force, and 0.2 m for –70 N force. **Point B**: the lever arm is 0.0 m for –50 N force, 0.6 m for –60 N force, and 0.45 m for –70 N force. **Point C**: the lever arm is 0.4 m for –50 N force, 0.2 m for –60 N force, and 0.0 m for –70 N force.

5. Using Figure 11.14 below list the direction of rotation (CW or CCW) caused by each force (–50, –60, –70) about points A, B, and C.

**Point A**: the –50 N force causes CW rotation –, the –60 N force causes no rotation, and the –70 N force causes CW rotation –. **Point B**: the –50 N force causes no rotation, the –60 N force causes CCW rotation +, and the –70 N force causes CCW rotation +. **Point C**: the –50 N force causes CW rotation –, the –60 N force causes CCW rotation +, and the –70 N force causes no rotation.

6. Compute the sum of the torques about point A.
$$\Sigma\tau = |F| \times l$$
$$\Sigma\tau = (-70N)(.2m) - (50N)(.6m)$$
$$\Sigma\tau = -44N \cdot m$$

7. Compute the sum of the torques about point B.
$$\Sigma\tau = |F| \times l$$
$$\Sigma\tau = +(60N)(.6m) + (70N)(.45m)$$
$$\Sigma\tau = +67.5N \cdot m$$

8. Compute the sum of the torques about point C.
$$\Sigma\tau = |F| \times l$$
$$\Sigma\tau = +(60N)(.2m) - (50N)(.4m)$$
$$\Sigma\tau = -8.0N \cdot m$$

9. The diver shown in Figure 11.15 is doing a backward two and one half rotation dive from a handstand position. Using Newton's law of angular acceleration, complete the free body diagram and determine the angular acceleration about the diver's center of mass.

$$\Sigma\tau = I\alpha$$

$$-130N(1.2m) - 1100N(.25m) = (10.2kg \cdot m^2)(\alpha)$$

$$\alpha = -42.25r / s^2$$

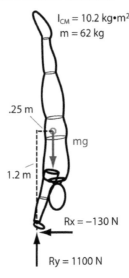

$I_{CM} = 10.2\ \text{kg} \cdot \text{m}^2$

$m = 62\ \text{kg}$

.25 m

mg

1.2 m

Rx = −130 N

Ry = 1100 N

## CHAPTER 13 ANSWERS

1. A diver in air has 70 kg·m²/s of angular momentum and a moment of inertia of 12 kg·m², what is his angular velocity? If the diver reduces his moment of inertia to 6 kg·m² while in the air how does this affect angular momentum and angular velocity?

$$L = I\omega$$

$$70kg \cdot m^2 / s = (12kg \cdot m^2)(\omega)$$

$$\omega = 5.83r / s$$

If the diver reduces his moment of inertia to 6 kg·m2 while in the air how does this affect angular momentum and angular velocity? Remember in the air angular momentum L is constant.

$$L = I\omega$$

$$70kg \cdot m^2 / s = (6kg \cdot m^2)(\omega)$$

$$\omega = 11.67r / s$$

2. An X games athlete applies a torque of 30 N·m for a time of t = 0.2 seconds while in a lay-out position with a moment of inertia I = 10 kg·m², his initial angular velocity was $\omega_i = 0.3$ r/s, what is his angular velocity after experiencing this torque? What angular impulse was applied?

$$\Sigma\tau(t) = I\omega_f - I\omega_i$$

$$(30N \cdot m)(.2s) = (10kg \cdot m^2)\omega_f - (10kg \cdot m^2)(.3r / s)$$

$$\omega_f = .9r / s$$

What angular impulse was applied? Angular impulse is torque × time.

$$\Sigma\tau(t) = (30N \cdot m)(.2s)$$
$$\Sigma\tau(t) = 6N \cdot m \cdot s$$

3. A diver with an initial angular velocity of −0.5 r/s and a moment of I = 7 kg·m² applies forces against the board which produce an angular impulse of 21 N·m·s, what is his final angular velocity after applying this impulse?

$$\Sigma\tau(t) = I\omega_f - I\omega_i$$

$$21N \cdot m \cdot s = (7kg \cdot m^2)\omega_f - (7kg \cdot m^2)(.5r/s)$$

$$\omega_f = 2.5r/s$$

4. Write the angular impulse – angular momentum equation and state the equation in words. What are the units for angular impulse and angular momentum? What is angular impulse? What is angular momentum? What is moment of inertia, and what are the units of moment of inertia?

$$\Sigma\tau(t) = I\omega_f - I\omega_i$$

In words this equation says "Angular Impulse is equal to the change in angular momentum", or "The sum of the torques x time is equal to the change in angular momentum". The units for angular impulse are N·m·s. The units for angular momentum are kg·m²/s. The units for moment of inertia are kg·m². Angular impulse is torque × time or the area under the torque – time graph. Angular momentum is the quantity of rotation a body possesses, it is a function of angular velocity × moment of inertia.

5. What is the Principle of Conservation of Angular Momentum? How do you create angular momentum? When can you change angular momentum? Draw a picture to show the relationship between angular momentum, angular velocity and moment of inertia for a diver in the air who goes from a layout to a tuck and back to a layout before entering the water. Principle of Conservation of Angular Momentum: When gravity is the only force operating (in the air) angular momentum is conserved (constant). You create angular momentum by applying an angular impulse or torque × time.

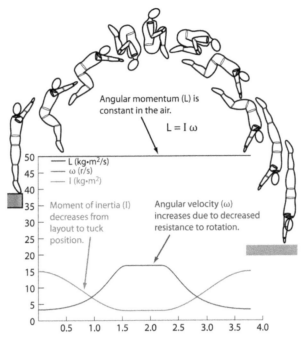

Angular momentum (L) is constant in the air.

$$L = I\,\omega$$

— L (kg·m²/s)
— ω (r/s)
— I (kg·m²)

Moment of inertia (I) decreases from layout to tuck position.

Angular velocity (ω) increases due to decreased resistance to rotation.

6. A torque –15 N·m is applied for t = 0.7 seconds on a rigid bar with an initial angular velocity of $\omega_i$ = +0.2 r/s and a moment of inertia of 10 kg·m², what was the final angular velocity after this torque was applied? What angular impulse was applied?

$$\Sigma\tau(t) = I\omega_f - I\omega_i$$

$$(-15N \cdot m)(.7s) = (10kg \cdot m^2)\omega_f - (10kg \cdot m^2)(.2r/s)$$

$$\omega_f = -0.85r/s$$

Angular impulse is (–15N·m)(0.7s) or –10.5N·m·s.

7. Download the Excel file Back Flip Practice File.xls and compute the following: angular velocity, angular momentum, angular impulse for all positive and negative areas of the torque – time curve, angular position. [Answer file available on course website.]

8. Download the Excel file: Rigid Bar.xls and compute the following angular velocity, angular momentum, angular impulse for all positive and negative areas of the torque – time curve, angular position. [Answer file available on course website.]

## CHAPTER 14 ANSWERS

1. Define the following terms: stress, strain, stiffness, compliance.
   Stress is force divided by area. It is a normalized variable in units of Pascal (Pa). Strain describes the change in shape of a material. Strain can be presented in absolute and relative units. Relative strain is a normalized variable. Stiffness is the slope of the stress – strain curve. Compliance is the inverse of stiffness. A compliant material provides less resistance to deformation then does a stiff material.

2. What is the difference between absolute and relative strain?

Absolute strain quantifies the actual amount of deformation of the material in units of mm, cm or m. Relative strain is a percentage change.

3. Draw a picture of each of the following types of loads: compression, tension, shear and torsion.

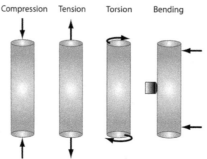

4. Draw a stress strain curve for a ductile, brittle and elastic material. On each graph identify the elastic region, yield point and plastic region.

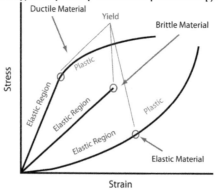

5. Define viscoelastic; explain the elastic response and the viscous response.

A viscoelastic material is an elastic material that contains some kind of fluid. The elastic response of a viscoelastic material is similar to a spring, it deforms and then returns almost all of the energy input to the system. The viscous response of a viscoelastic material will absorb some of the energy input to the system and in addition the viscous component is velocity dependent. When loaded slowly the viscous response is compliant and it becomes progressively stiffer with increasing velocity of loading. Viscoelastic materials exhibit creep, force relaxation and a hysteresis.

6. Describe the difference between a creep response and a force relaxation response when loading a viscoelastic material.

   In the creep response a constant force is applied to the material. When the force is initially applied the material will deform rapidly, after the initial rapid deformation the material will slowly continue to deform. The later slow deformation is creep.

In force relaxation the material is deformed to a fixed length, when it is initially deformed the material is stiff, indicating that the resistance to deformation is high. When left at the fixed length the internal resistance declines somewhat. This decline in internal resistance is force relaxation.

## CHAPTER 15 ANSWERS

1.  Describe the structural similarities and differences between cortical and cancellous bone.
    Cortical and cancellous tissue both contain osteocytes with Haversian and Volkmann's canals. Both cortical and cancellous tissue are arranged in lamellar layers, but the layers are different. In cortical tissue the lamellar layers form columns that are wrapped inside columns running in the direction of the long axis of the bone. In cancellous tissue the lamellar layers form trabecular plates or trabecular rods that are aligned according to the principal direction of loading.

2.  What is a basic multicellular unit? Describe its role in remodeling.
    A basic multicellular units (BMUs). BMUs include osteocytes, osteoclasts, osteoblasts and bone lining cells (Frost, 1987). Bone lining cells, osteoblasts and osteoclasts are located on the bone surfaces, and the osteocytes are located in the bone matrix. The BMUs work in concert to remodel tissue. Strains in the tissue produce micro-cracks in the canaliculi. The osteocytes whose canaliculi are cracked under a programmed death. Osteoclasts dig a hole in the tissue and then osteoblasts lay down new tissue.

3.  What is muscular preactivation?
    The neuromuscular system is relatively slow. It requires 80 – 200 ms to respond to a stimulus. In fast events such as landing or running the muscular system is activated prior to the actual event. About 100 – 200 ms before landing the ankle, knee and hip joint muscles are activated to prepare for the collision with the ground. The amount of preactivation of each muscle is based upon previous experience. If something unexpected happens, there is no large safety factor to protect the system.

4.  What is the minimum time for the neuromuscular system to respond to a stimulus? What factors affect the minimum time for the neuromuscular system to respond to a stimulus?
    The minimum time for the neuromuscular system to respond to a stimulus is 80 – 200 ms. This includes: reflex time, muscle force time and time to over come momentum.

5.  What methods were described to determine the quality and properties of bone?
    Material testing systems are used to determine the stress – strain properties of bone in compression, tension, shear and torsion. Dexa machines are used to determine the bone mineral density.

6.  Describe the role of cartilage in attenuating forces.

Cartilage and synovial fluid are very effective in force attenuation. Under low magnitude and low rate of force loading the cartilage is not stiff and it passes the load onto the bone, effectively generating a stimulus for adaptive remodeling. Under high rates and high magnitude of loading the cartilage is stiff and it protects the bone from high frequency shock wave that would damage the tissue and cause an excessive number of micro-cracks. Thus cartilage and synovial fluid can be described as an adjustable shock absorber system.

7. Describe the long term effect of impact forces on joints as observed by Radin.
Repeated impacts to joints eventually cause the cartilage to become thin or non existent. Thinning of the cartilage causes the cortical bone to thicken and the cancellous bone to become stiff and brittle. These series of events lead to pain during standing and walking in the joints.

# CHAPTER 16 ANSWERS

1. What causes disinhibition in the actin myosin to enable the formation of a crossbridge?
Under resting conditions the two regulatory proteins troponin and tropomyosin inhibit myosin from forming a crossbridge with actin. Once $Ca^{++}$ is released from the sarcoplasmic reticulum it binds with troponin. The binding of $Ca^{++}$ with troponin causes a change in the configuration of the actin filament. In the resting state tropomyosin covers the actin binding site and prevents (inhibits) the formation of a crossbridge. The change in configuration of the actin filament exposes the actin binding site and since myosin has a high affinity to bond with actin a crossbridge between actin and myosin is formed. The configurational change of actin removes the normal inhibition actin – myosin bonding that is regulated by troponin and tropomyosin and is referred to as disinhibition.

2. What structures are responsible for increased muscle stiffness with increasing force?
The increased muscle stiffness is attributed to elastic strain of myosin, actin, desmin, costamere, and titin.

3. Who are the two scientists who were instrumental in the development of the sliding filament theory? Andrew Huxley and Hugh Huxley.

4. What is the range of force generated in a single concentric actin – myosin crossbridge from the slowest to the fastest velocity of shortening?
The actual amount of force generated in a concentric actin – myosin crossbridge cycle will vary according to the velocity of muscle shortening. At slow velocities of shortening the force may be as high as 6 pN and at fast velocities of shortening the force may be as low as 4 pN.

5. How does the velocity of lengthening affect the force generated in a single eccentric actin – myosin crossbridge cycle?

A single actomyosin crossbridge in an eccentric contraction generates 6 pN of force. The total force generated is 10-20 pN. The additional 4-14 pN of force comes from elastic strain of crossbridges and cytoskeletal tissues.

6. List the steps from the stimulation of a nerve to the production of tension.
    1. A stimulus is propagated down the alpha motor neuron.
    2. Acetylcholine (Ach) is released from the endplate and crosses the synapse.
    3. Ach causes $Na^+$ and $K^+$ channels to open up on the sarcolemma.
    4. $Na^+$ flows into the cell and $K^+$ flow out of the cell, generating a muscle fiber action potential.
    5. $Na^+$ spreads downward into the T-tubule system causing $Ca^{++}$ to be released from the sarcoplasmic reticulum.
    6. $Ca^{++}$ binds with troponin, a change in configuration of actin that exposes the actin binding site.
    7. A crossbridge is formed between actin and myosin.
    8. The ATP in the myosin head is downgraded to ADP + Pi.
    9. Once the Pi is released the myosin head is tightly bound to actin.
    10. The myosin arm does work on actin and tension is generated.

11. What structure on the actin filament does the myosin head attach to in order to form a crossbridge?
    The myosin arm forms a crossbridge by attaching to the binding site on actin.

12. What event causes the configurational change in the actin filament that exposes the actin binding site?
    The actin binding site is exposed after $Ca^{++}$ binds with troponin.

13. What makes the muscle fiber lengthen in an eccentric contraction?
    An external force causes the fiber to lengthen. The actin – myosin crossbridge always attempts to shorten the fiber, but the external force causes it to lengthen.

# CHAPTER 17 ANSWERS

1. Draw and explain the force-length curve for a sarcomere.

There is an inverted U relationship between fiber length and fiber force in a sarcomere. At short lengths the force is low because of overlapping or interference of actin-myosin crossbridges. At optimal lengths more crossbridges are formed which then generates the highest level of force. At longer lengths some of the crossbridges are not able reach each other so force declines.

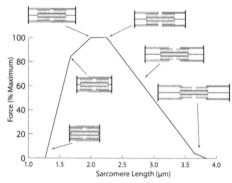

2. Draw and explain the force-velocity curve.

Eccentric force > isometric force > concentric force. For concentric contractions, an increase in the velocity of shortening results in a decrease in force because of increased turnover of crossbridges and reduced crossbridge force 4-6pN. In eccentric contractions an increase in the velocity of lengthening results in an increase in force. Of this, 6 pN comes from crossbridges and 4-14 pN comes from elastic strain of crossbridges and cytoskeletal structures.

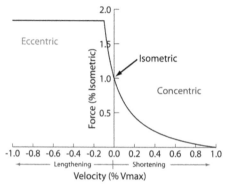

The eccentric force reaches a plateau where further increases in the velocity of stretch yield a constant eccentric force.

3. Define a stretch-shorten cycle and give an example of a movement that utilizes a stretch-shorten cycle.
   A stretch-shorten cycle is defined as an eccentric contraction followed by an immediate concentric contraction. A vertical jump or a fast bench press are both examples of a stretch-shorten cycle.

4. List four explanations for the additional work that is done during the concentric phase of a stretch-shorten cycle.
   1. The eccentric phase stores elastic energy which is returned during the shortening phase.
   2. The eccentric phase causes a stretch reflex which increases the neural drive to the muscle.
   3. The prior activation during the eccentric phases allows the muscle to start the concentric phase at a higher force level.
   4. Two joint muscles allow provide increased efficiency allowing the muscle to work at a lower velocity and a higher force.

5. List the anatomical structures that are associated with each of the following components of the three component muscle model: parallel elastic component, contractile component, and series elastic component.
   The contractile component (CC) represents active crossbridges generating tension (the en-

gine of the muscle machine) and it also contains a dashpot to explain the viscous damping response of muscle. The series elastic component (SEC) represents all anatomical structure in series (in line with) the force generating contractile component. The anatomical structures included in the series elastic component are the tendon, titin, desmin and the bending or stretching of the myosin arms. The third component the parallel elastic component (PEC) includes passive elastic structures: tendon, passive crossbridges, endomysium, perimysium, epimysium, titin and desmin.

## CHAPTER 18 ANSWERS

1. Describe the relationship between muscle fiber types (ST, INT, FT), firing rates, twitch forces and fatigability.
   Slow twitch fibers generate lower force with a longer time to peak twitch force, the firing rates range from 5–60 Hz and they are highly resistant to fatigue. Intermediate fibers generate higher force with a shorter time to peak twitch force then ST fibers, the firing rates range from 10–80 Hz and they are somewhat fatigue resistant in that they can be driven continuously for 4–6 minutes. Fast twitch fibers generate the highest twitch force with the shortest time to peak twitch force, they have firing rates ranges from 10–120 Hz, and they highly fatigable.

2. Draw a picture of a motor unit and show muscle fiber action potentials being propagated each direction along the muscle fiber from the terminal branches of the nerve.

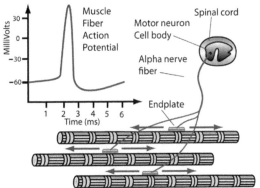

3. What is the definition of the electromyogram?
   The EMG signal is defined as the sum of all of the muscle fiber action potentials from all of the active motor units that pass thru the electrode recording zone.

4. What factors affect the shape of the electromyogram?
   The shape of the EMG signal is affect by the direction the muscle fiber action potential travels thru the recording zone, the distance along the fiber, and the distance from the fiber to the electrodes.

5. What is the difference between a muscle fiber action potential and a motor unit action potential?

A muscle fiber action potential is the potential from a single muscle fiber that is propagated along the muscle fiber. A motor unit action potential is the sum of all the muscle fiber action potentials from a single motor unit that pass thru the recording zone.

6. Give an example of how synchronization of motor units can cause amplification and cancellation of the EMG signal.
   Synchronization primarily causes cancellation of motor unit action potentials. Occasionally synchronization will cause amplification of motor unit action potentials.

7. Draw a picture to show how stimulation or firing rate affects the force – time curve.

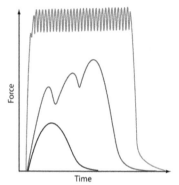

Stimulation or firing rate increases causes a stair-stepping effect of force.

8. Describe the size principle.
   Motor units are recruited by size in this order ST, INT, FT. Size can refer to the size of the force signal, size of the electrical signal and the thickness of the alpha nerve fibers.

9. Draw a picture to show EMG recruitment and firing rate changes during a ramp contraction.

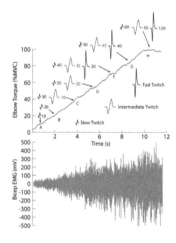

10. How would motor unit recruitment and firing differ when comparing a ramp to a ballistic contraction?
    In a ramp contraction there would be orderly recruitment ST, INT, FT with some time in-

terval between recruitment. Also in a ramp the motor unit firing rates would be low when recruited and they would increase the rates slowly. In a ballistic contraction they would be recruited almost simultaneously and each motor unit would be firing at a high rate at the outset.

11. How is recruitment threshold related to the size principle?
In a ramp contraction motor units are recruited according to size. They are turned on at a specific force threshold and derecruited when the force goes back below this same force level.

12. A subject generates a ramp contraction in which he/she slowly increases the force from 0 – 45% MVC. Describe the expected recruitment and firing rate patterns.
The contraction would be generated using only ST and INT fibers. Each unit would slowly increase its firing rate. New units would enter the contraction firing at a slower rate then currently active motor units.

13. A subject generates a ballistic contraction in which he/she increases the force from 0 – 45% MVC as rapidly as possible. Describe the expected recruitment and firing rate patterns.
ST and INT fibers would fire at relatively high rates. Since it is ballistic FT fibers would fire at a high rate initially and then shut off once the force level of 45% is reached.

14. What is electromechanical delay? What factors are attributed to electromechanical delay? What is the primary factor that causes electromechanical delay?
Electromechanical delay is the delay between the start of EMG and the start of force. It is attributed to the spread of $Na^+$ into the muscle cell, the release of $CA^{++}$, the formation of a crossbridge, taking up slack in titin and the tendon. Almost all of the EMD time is actually due to taking up slack in the tendon.

## CHAPTER 19 ANSWERS

1. Describe a stretch reflex.
In a stretch reflex the muscle is pulled and the amount and rate of stretch exceeds the currently set sensitivity of the muscle spindle. This causes the Ia neuron to send impulses the spinal cord. These impulses cause: 1) the muscle that was stretched and its synergist to receive alpha stimulation so they shorten, 2) the antagonists of the muscle that was stretched receive alpha inhibition so they relax, and 3) the Ia nerve fires the gamma neuron which causes the specialized muscle tissue on the muscle spindle to reset the spindle length so that it can now monitor the muscle at its new length.

2. List the afferent and efferent fibers on a muscle spindle and a Golgi tendon organ.
The muscle spindle has two afferent fibers: II and Ia and one efferent fiber gamma. The Golgi tendon organ only has one afferent fiber Ib and no efferent fibers.

3. What is the difference between a stretch reflex and an inverse stretch reflex.
In a stretch reflex the muscle is pulled and the response causing the muscle to shorten. In an inverse stretch reflex (as in stretching a muscle for flexibility) the process begins with a cortical command causing the muscle to be stretched to relax and the spindle of the muscle to be stretch to relax (loosen). The goal in an inverse stretch reflex is the stretch the muscle without causing a stretch reflex.

4. Give an example of a motion that uses feedback and feedforward control.
Motions that are performed slow > 100 – 200 ms can use feedback to adjust errors. Balancing on a balance beam or slowly walking across a balance beam would be an example of movements that could use feedback to adjust errors. Movements that occur quickly in less than 100 ms can not use feedback so they must rely on feedforward control. A catcher who catches a fastball that has been deflected off from the batter's bat can not use feedback to adjust the position of the glove. This motion happens without feedforward control also since there is not time for the catcher to plan what is going to happen. An example of feedforward control would be the batters motion to hit a ball. Once the batter sees the ball leave the pitchers hand he or she makes a guess where the ball will be and then attempts the place the bat in the projected (guessed) position.

## CHAPTER 20 ANSWERS

1. Describe sensor information that the brain uses to organize a response necessary to maintain balance.
To maintain balance the brain must keep the center of mass over the base of support. This is accomplished by monitoring somatosensory, vestibular and visual signals. The brain receives signal from all three, it must decide if there are error or inaccuracies in any input signals and then organize a motor response to maintain balance.

2. An individual is standing on a moving platform with his eyes closed describe the role of vestibular and somatosensory information available to the brain to organize postural control responses.
Since the eyes are closed vision is not available. If the platform is traveling on a smooth surface at a constant speed there will be no accelerations of the head, thus very little information from the vestibular sensors. If the platform accelerates (slows down or speeds up) or hits a bump then vestibular signals will provide information to the brain. Somatosensory signals will be available to the brain throughout.

# APPENDIX B EQUATION SHEET

$$V = \frac{P_f - P_i}{\Delta t} \qquad \Delta P = \int V dt \qquad P_f = \int (V \times dt) + P_i$$

$$A = \frac{V_f - V_i}{\Delta t} \qquad \Delta V = \int A dt \qquad V_f = \int (A \times dt) + V_i$$

$$x = r \cos\theta$$

$$y = r \sin\theta$$

$$r = \sqrt{x^2 + y^2}$$

$$Vy_f = Vy_i + at \qquad\qquad x = Vx(t)$$

$$\theta = tan^{-1}\left(\frac{y}{x}\right)$$

$$(Vy_f)^2 = (Vy_i)^2 + 2a(y_f - y_i)$$

$$y_f = y_i + Vy_i(t) + \frac{1}{2}at^2 \qquad\qquad X_{CM} = \frac{m_1 x_1 + m_2 x_2 + m_3 x_3 + \cdots}{m_1 + m_2 + m_3 + \cdots}$$

$$Y_{CM} = \frac{m_1 y_1 + m_2 y_2 + m_3 y_3 + \cdots}{m_1 + m_2 + m_3 + \cdots}$$

$$v = \omega r$$

$$\omega = \frac{\theta_f - \theta_i}{\Delta t} \qquad a_T = \alpha r \qquad a_C = r\omega^2 \qquad a_C = \frac{V^2}{r}$$

$$\alpha = \frac{\omega_f - \omega_i}{\Delta t} \qquad a_T = \frac{V_f - V_i}{t} \qquad F_C = mr\omega^2 \qquad F_C = m\frac{V^2}{r}$$

$$rad = \frac{deg}{57.29} \qquad kg = \frac{Lbs}{2.2046} \qquad g = -9.8\,\text{m/s}^2 \qquad Wt = mg \qquad \theta = \frac{s}{r}$$

$$\Sigma Fx = ma_x \qquad \Sigma Fy = ma_y \qquad \tau = \pm|F|\times l$$

$$\Sigma\tau = I\alpha$$

$$V_A = \frac{m_B u_B}{m_A} \qquad m_A u_A + m_B u_B = (m_A + m_B)V$$

$$\pm|Rx|\times l \pm|Ry|\times l = I_{CM}\alpha_{CM}$$

$$V_B = \frac{m_A u_A}{m_B} \qquad J = \overline{F}\Delta t \qquad J = \int F\Delta t$$

$$p = mv$$

$$\overline{F}(\Delta t) = mV_f - mV_i$$

$$L = I\omega$$

$$\overline{R}x(\Delta t) = mVx_f - mVx_i \qquad \overline{R}y(t) + mg(t) = mVy_f - mVy_i \qquad \Sigma\tau(t) = I\omega_f - I\omega_i$$

$$\int_{t_0}^{t_1} Rx\, dt = mVx_f - mVx_i \qquad \int_{t_0}^{t_1}(Ry + mg)\, dt = mVy_f - mVy_i \qquad \int_{t_0}^{t_1}\tau\, dt = I\omega_f - I\omega_i$$

$$W = F(y_f - y_i) \qquad KE = \frac{1}{2}mVy^2 \qquad PE = mgh \qquad g = +9.8$$

$$W_R = \tau(\theta_f - \theta_i) \qquad\qquad E = KE + PE$$

$$KE_R = \frac{1}{2}I\omega^2 \qquad\qquad h = \frac{E}{mg}$$

$$W = \frac{1}{2}m(Vy_f)^2 - \frac{1}{2}m(Vy_i)^2 + mgh_{y_f - y_i} \qquad P_R = \frac{W_R}{t} \qquad P = \frac{W}{t}$$

$$P_R = \tau \times \omega \qquad P = F \times V$$

$$W = \int_{t_0}^{t_1} P \times \Delta t \qquad W = \int_{S_0}^{S_1} F \times \Delta s \qquad W_R = \int_{\theta_0}^{\theta_1}\tau \times \Delta\theta$$

# Index